Modeling and Control of Energy Conversion during Underground Coal Gasification Process

Modeling and Control of Energy Conversion during Underground Coal Gasification Process

Editors

Marek Laciak
Ján Kačur
Milan Durdán

MDPI • Basel • Beijing • Wuhan • Barcelona • Belgrade • Manchester • Tokyo • Cluj • Tianjin

MDPI

Editors
Marek Laciak
Technical University of Kosice
Slovakia

Ján Kačur
Technical University of Kosice
Slovakia

Milan Durdán
Technical University of Kosice
Slovakia

Editorial Office
MDPI
St. Alban-Anlage 66
4052 Basel, Switzerland

This is a reprint of articles from the Special Issue published online in the open access journal *Energies* (ISSN 1996-1073) (available at: https://www.mdpi.com/journal/energies/special_issues/Modeling_and_Control_of_Energy_Conversion_during_UCG_Process).

For citation purposes, cite each article independently as indicated on the article page online and as indicated below:

LastName, A.A.; LastName, B.B.; LastName, C.C. Article Title. *Journal Name* **Year**, *Volume Number*, Page Range.

ISBN 978-3-0365-6011-3 (Hbk)
ISBN 978-3-0365-6012-0 (PDF)

Contents

About the Editors

Marek Laciak

Marek Laciak is a Professor at the Institute of Control and Informatization of Production Processes, Faculty BERG at the Technical University of Kosice. In 2022, he defended his inaugural thesis and received a professorship in automation. He is the author and co-author of three scientific monographs and 150 papers published in reputed journals and conferences. He has 307 citations with an h-index of 10, according to WoS (https://publons.com/wos-op/researcher/2694333/marek-laciak/). He has participated in 21 national and international research projects, in 3 of which he was the principal investigator. He has been the supervisor for 63 bachelor's and diploma theses and 2 doctoral theses. He is a member of the program committee of the International Carpathian Control Conference, the committee of the Slovak Association of Applied Cybernetics and Informatics (SSAKI), and the committee of the Czech & Slovak Society for Simulation (CSSS). His research interests include: modeling, monitoring, and controlling technological processes; indirect measurement systems; the optimization of processes; and underground coal gasification.

Ján Kačur

Ján Kačur is an Associate Professor at the Institute of Control and Informatization of Production Processes, Faculty BERG at the Technical University of Kosice. From a professional point of view, he is dedicated to automation and the design of control systems. He is also devoted to mathematical modeling based on machine learning. His research interests focus on underground coal gasification and steelmaking. He is the author and co-author of 61 publications indexed by WoS (see https://publons.com/wos-op/researcher/2678181/jan-kacur/) and 1 monograph. He has 305 citations with an h-index of 10, according to WoS. He has participated in 18 national and international research projects. He is a member of the program committee of the International Carpathian Control Conference and a member of the Slovak Association of Applied Cybernetics and Informatics (SSAKI).

Milan Durdán

Milan Durdán is an Associate Professor at the Institute of Control and Informatization of Production Processes, Faculty BERG at the Technical University of Kosice. He completed his habilitation thesis on earth resources extracting and processing in 2018 at the Technical University in Košice. He is the author of a monograph entitled Annealing process control of steel coils, which focuses on the proposal of models for calculating the temperature inside an annealed coil. He is also the author of three teaching texts focused on modeling and optimizing the processes of the extraction and processing of raw materials. He has 56 publications and 299 citations with an h-index of 10, according to WoS (https://publons.com/researcher/2678237/milan-durdan/). He has participated in 16 national and international research projects, in 1 of which he was the principal investigator. He has been the supervisor for 9 bachelor's and 16 diploma theses in the field of mathematical and computer modeling, and the monitoring extraction and processing of raw material processes. In his professional life, he is focused on mathematical modeling and optimization of raw material extraction and processing processes. The applications of his research activity are primarily focused on modeling, optimization, and monitoring systems and their use in the field of technological process control. He has been a member of the Slovak Association of Applied Cybernetics and Informatics (SSAKI) since 2005.

Preface to "Modeling and Control of Energy Conversion during Underground Coal Gasification Process"

The underground coal gasification (UCG) technology is an unconventional method of coal mining, and its approaches represent new scientific knowledge. This continually evolving technology is a large energy source that can be obtained at a lower cost than convection mining and is also safer. The UCG process transforms the coal's energy into the gas produced (i.e., syngas) during the underground coal gasification process in the coal seam.

For successful energy conversion, i.e., obtaining the syngas with a higher calorific value, it is essential to develop new modeling methods and control this process. Modeling methods will make it possible to identify the individual stages of the UCG process more precisely and, thus, improve the knowledge of this process. Various UCG models, e.g., models based on CFD, machine models for syngas composition, or temperature prediction, were recently investigated. By synthesizing suitable algorithms and mathematical models, it is possible to obtain simulation models of the process. These simulation models can be used to design the control systems and the optimal setting of process parameters. In the control of UCG, the significant issue is to ensure sufficient underground temperature and the calorific value of syngas. This issue can be solved by advanced control methods and optimization of operating variables.

The present book contains nine articles that were accepted and published in the Special Issue *"Modeling and Control of Energy Conversion during Underground Coal Gasification Process"* of the MDPI Energies journal. This book focuses on the energy conversion processes in underground coal gasification (UCG), as well as on the modeling and control of this process.

Marek Laciak, Ján Kačur, and Milan Durdán
Editors

energies

MDPI

Editorial

Modeling and Control of Energy Conversion during Underground Coal Gasification Process

Marek Laciak *, Ján Kačur and Milan Durdán

Institute of Control and Informatization of Production Processes, Faculty BERG, Technical University of Košice, Němcovej 3, 042 00 Košice, Slovakia; jan.kacur@tuke.sk (J.K.); milan.durdan@tuke.sk (M.D.)
* Correspondence: marek.laciak@tuke.sk; Tel.: +421-55-6025175

1. Introduction

The underground coal gasification (UCG) technology is an unconventional method of coal mining, and its approaches represent new scientific knowledge. This continually evolving technology is a large energy source that can be obtained at a lower cost than convection mining and is also safer. The UCG process transforms the coal's energy into the gas produced (i.e., syngas). The coal is converted into syngas during the underground coal gasification process in the coal seam (i.e., in situ). The gas is produced and extracted by a well drilled into the gasified coal seam. The injection well is used to inject oxidants, i.e., air, oxygen, vapor, or their mixture. The injection well also serves to ignite the coal seam at the beginning of the UCG process. Production wells are used to transport the product gas to the Earth's surface (Figure 1) [1,2]. High-pressure gasification is carried out at a temperature of 700–900 °C, but under certain conditions, a temperature of up to 1500 °C can be achieved [2,3]. The coal is decomposed and mainly produces carbon dioxide (i.e., CO_2), hydrogen (i.e., H_2), carbon monoxide (i.e., CO), small amounts of methane (i.e., CH_4), and hydrogen sulfide (i.e., H_2S) in the UCG process [3,4].

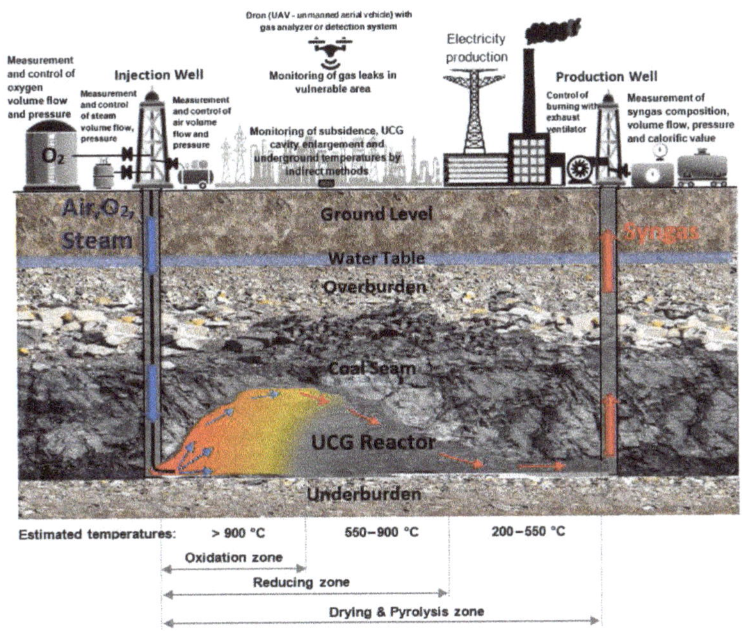

Figure 1. Principle of the UCG process [5].

Citation: Laciak, M.; Kačur, J.; Durdán, M. Modeling and Control of Energy Conversion during Underground Coal Gasification Process. *Energies* **2022**, *15*, 2494. https://doi.org/10.3390/en15072494

Received: 18 February 2022
Accepted: 25 March 2022
Published: 28 March 2022

For successful energy conversion, i.e., obtaining the syngas with a higher calorific value, it is essential to develop new modeling methods and control this process. Modeling methods will make it possible to identify the individual stages of the UCG process more precisely, and thus improve the knowledge of this process. Recently, various UCG models, e.g., models based on CFD, machine models for syngas composition, or temperature prediction, were investigated. By synthesizing suitable algorithms and mathematical models, it is possible to obtain simulation models of the process. These simulation models can be used to design the control systems and the optimal setting of process parameters. In the control of UCG, the significant issue is to ensure sufficient underground temperature and the calorific value of syngas. This issue can be solved by advanced control methods and optimization of operating variables.

This Special Issue (SI) of Energies journal focused on the energy conversion processes in underground coal gasification and the modeling and control of this process. A brief review of the articles published in the SI "Modeling and Control of Energy Conversion during Underground Coal Gasification Process" will be presented in the next section.

2. Special Issue Articles' Short Review

Articles published in Energies as a part of this Special Issue can be divided into the three thematic parts of research in the field of underground coal gasification technology. The first part is the impact of technology on the environment, the second is research (studies) the coal areas and coal properties for UCG technology, and the third is the monitoring, modeling, and control of UCG processes.

This study [6] provides an overview of the systematic methods of the in situ coal gasification process. In the paper, it has been presented the model of the porous structure of coal and the gas movement taking place in the carbon matrix, which is part of the bed. The experimental tests were carried out with the use of air forced through the nozzle in the form of a gas stream spreading in many directions in a porous bed under bubbling conditions. The gas flow resistance coefficient was determined as a function of the Reynolds number in relation to the diameter of the gas flow nozzle. The evaluation showed that the models available in the literature have a limited scope of application to skeletal media, characterized by a significant internal structure of the porous material. The test results show that under bubbling conditions, it is possible to accurately assess the gas permeability, which makes it possible to comprehensively assess the properties of the porous material in terms of the process for UCG technology.

The main purpose of the study described in the article [7] was a qualitative and quantitative characterization of the UCG wastewater produced during four different UCG experiments. The experiments were carried out in an ex situ UCG installation located in the Clean Coal Technology Centre of the Central Mining Institute (Mikołów, Poland). One of the most important issues during UCG process is wastewater production and treatment. Condensed gasification wastewater is contaminated by many hazardous compounds. The composition of the generated UCG-derived wastewater may vary depending on the type of gasified coal and conditions of the gasification process. The conducted studies revealed significant relationships between the physicochemical composition of the wastewater and the coal properties, as well as the gasification pressure. The strongest impact is noticeable in the case of organic pollutants, especially phenols, BTEX and PAHs. The most abundant group of pollutants were phenols. The experimental studies have shown that concentrations of phenols, BTEX, and PAHs decrease with increasing pressure.

The conducted research has shown that UCG wastewater contains many hazardous pollutants and requires the selection of an appropriate treatment method, for example, for coking wastewater. The presented results in the paper [7] can help in the development of an appropriate UCG wastewater treatment strategy depending on the coal used and gasification parameters.

Article [8] presents the results of laboratory tests regarding the influence of high temperatures on changes in the strength and structural parameters of rocks that are present in the immediate vicinity of a gasification channel. Sandstone and claystone samples were heated at 300 °C, 600 °C, 900 °C, and 1200 °C. Additionally, the heated samples were placed in water for 24 h. Strength tests regarding sandstone and claystone were carried out in a hydraulic press at the laboratory of the Faculty of Civil Engineering and Resource Management at the University of Science and Technology in Krakow.

The results of the laboratory tests were used in the numerical simulation using RS2 software. The main goal of modeling was to determine the extent of the rock destruction zone around the gasification channel for dry and wet rock masses. On the basis of the obtained results, it was found that the extent of rock destruction, both in the roof and in the floor, is greater by several percent for a wet rock mass. For the first time, this research presents the effect of water on heated rock samples in terms of the underground coal gasification process. The results of laboratory tests and numerical simulations clearly indicate a reduction in strength, deformation, and structural parameters for the temperature of 1200 °C. Based on numerical results of the research, it can be concluded that, for the width of the gasification channel equal to 10, 20, and 30 m, the maximum extent of rock destruction for dry rock mass does not exceed 5, 10, and 15 m, but an increase in the extent of rock destruction occurs for wet rock mass [8].

Another paper [9] of this Special Issue was focused on the effect of lignite properties on its suitability for the implementation of UCG process. Two experimental simulations of UCG processes, using large bulk samples of lignites, were conducted in a surface laboratory setup. Two different lignite samples were used for the oxygen-blown experiments, i.e., "Velenje" meta-lignite (Slovenia) and "Oltenia" ortho-lignite (Romania). The average moisture content of the samples was 31.6 wt.% and 45.6 wt.% for the Velenje and Oltenia samples, respectively. The main aim of the study was to assess the suitability of the tested lignites for the UCG process. The gas composition and its production rates, as well as the temperatures in the artificial seams, were continuously monitored during the experiments. The average calorific value of gas produced during the Velenje lignite experiment (6.4 MJ/Nm^3) was much higher compared to the result obtained for the experiment with Oltenia lignite (4.8 MJ/Nm^3). The Velenje lignite test was also characterized by significantly higher energy efficiency, i.e., 44.6%, compared to the gasification of Oltenia lignite (33.4%). The gasification experiments carried out showed that the physicochemical properties of the lignite used considerably affect the in situ gasification process. The results of the study indicate that underground gasification may be a feasible option for the extraction of lignite deposits, especially in the case of Velenje lignite, which was characterized by a relatively higher calorific value, and lower moisture and ash content.

The research presented in paper [10] is based on experimental studies UCG process as in most articles of this Special Issue. The experimental equipment is designed and patented at Dnipro University of Technology, and manufactured by Naftomash RMA under financial support of the Ministry of Education and Science of Ukraine. A gas generator model consists of four systems:

- An experimental stand;
- A system of supply of separated and mixed blow mixture;
- A gas outlet system;
- A system of control and measuring equipment (temperature control and control of input and output gas mixtures).

This paper represents the results of experimental studies of physical modeling of the underground coal gasification process in terms of implementation of design and technological solutions aimed at intensification of a gasification process of thin coal seams. A series of experimental studies were performed in terms of a stand unit with the provided criteria of similarity to field conditions as well as the kinetics of thermochemical processes occurring within a gas generator. Hard coal (high volatile bituminous coal) was selected as the raw material to be gasified, as that coal grade prevails in the Ukrainian energy balance, since it is represented by rather great reserves. Five blow types were tested during the research (air, air–steam, oxygen–steam, oxygen-enriched, and carbon dioxide and oxygen). As a result, the effect of the tightness of a gas generator on the quantitative and qualitative parameters of coal gasification while varying the blow by reagents and changing the pressure in a reaction channel has been identified. Special attention was paid to the design solutions involving blow supply immediately into the combustion face of a gas generator. The experimental results demonstrate maximum efficiency of the applied gas generator design involving flexible pipelines and activator in the reaction channel and a blow direction onto the reaction channel face combined with blow stream reversing, which will make it possible to improve caloricity of the generator gas up to 18% (i.e., from 8.4 to 12.8 MJ/m^3 depending upon a blow type). Consideration of the obtained results of physical modeling can be used with sufficient accuracy to establish modern enterprises based on the underground coal seam gasification; this will help develop more efficiently the substandard coal reserves to generate heat energy as well as power-producing and chemical raw material. The research conclusions can provide technical references for developing a new generation of UCG technology [10].

Paper [11] researches the possibility of the model's utilization for temperature prediction in UCG process. Within experimental research, several regression models were proposed that differed in their structures, i.e., the number and type of selected controllable variables as independent variables. The goal was to find an optimal regression model structure, where the underground temperature is predicted with the greatest possible accuracy. The regression model structure proposal was realized on data obtained from two laboratory measurements realized in the ex situ UCG reactor. These experiments differed by the volume of gasified coal and thus also in the duration of the experiment. The proposal of regression coefficients was performed on the data from the first experiment, but the verification of the proposed regression models was performed mainly on the data from the second experiment. The results of temperature models were evaluated using the multiple coefficient of determination R^2. The values of this coefficient were during the verification of models on the second experiment lower than their values in the first experiment. The maximum value of the multiple coefficient of determination was reached at the temperature *T8* (R^2 = 0.87). The behavior of the measured and predicted temperature *T8* by the model is shown in Figure 2.

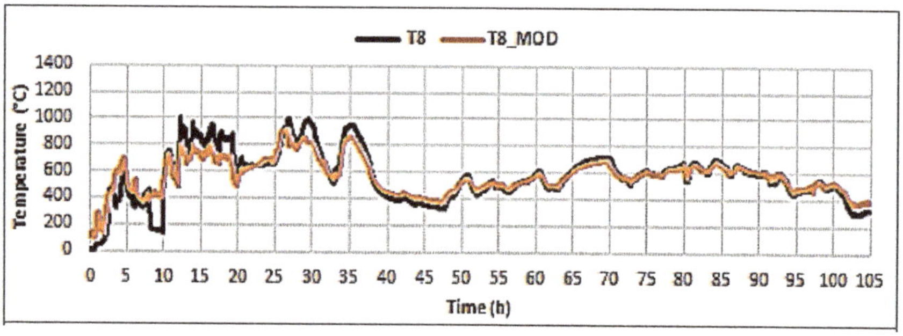

Figure 2. The measured (T) and modeled (T_MOD) temperature behavior [11].

The proposed models should contribute to developing a methodology for predicting temperatures in a gasified coal seam. Improving the prediction of these temperatures with higher accuracy makes it possible to identify places in the coal seam where coal to gas is transformed, and the underground cavity is formed. In addition, the prediction of coal seam temperatures allows the development of methods to control the UCG process based on modeled temperatures in the coal seam [11].

Paper [12] presents an experimental study of optimization of operating variables (airflow, oxygen flow, and syngas exhaust) during gasification in ex situ reactor. Optimization aims to maximize syngas calorific value. The proposed optimization algorithm was based on a simple gradient method that optimizes by experimental way of operating variables. The novelty and originality of the proposed solution of UCG control rests on a model-free approach. This approach of automated control of UCG process to maximize calorific value has not been investigated to date. Most of the research in the world focuses on the model-based stabilization of calorific value and mathematical modeling of UCG processes. However, research is lacking in improving the direct automated control of the UCG process that is often controlled only blindly.

The proposed control algorithm has been implemented on a PLC; it does not require a process model, and only online measured process data are needed. The algorithm, in four tests, was able to increase the calorific value by optimizing the operating variables. Better algorithm performance, i.e., higher syngas calorific value, was achieved by optimizing three operating variables, i.e., when additional oxygen flow was optimized [12].

Article [13] investigated the possibility of using coal in situ, using UCG technology. The authors of the paper focused on verified geological, hydrogeological, and tectonic information about the selected brown coal deposit in Slovakia. Based on the analysis and obtained information, possible adverse factors were evaluated. These factors affect the rock environment around the underground generator by UCG activity. The article also draws attention to the possible impact of pollution, taking into account the geological, hydrogeological, and tectonic conditions in the selected locality. Attention was focused on pollution from UCG process after experimental gasification, taking into account the amounts of gasified coal based on analyses of tar.

3. Conclusions

The articles presented in the Special Issue "Modeling and Control of Energy Conversion during Underground Coal Gasification Process" reflect current research trends in the field of underground coal gasification. Most of the results reported in the articles are based on experiments performed in laboratory conditions (ex situ reactor). These experiments are significant for verifying the technology and its implementation in real conditions. Based on experiments in laboratory conditions, the following is possible:

- Assess the suitability of the coal deposit for UCG technology;
- Assess the impact of UCG technology on the environment;
- Design and verify mathematical models that can use for the control and optimization of the UCG process.

We firmly believe that the articles presented in the Special Issue will help in the advancement of UCG technology and will be an inspiration for experts in this field.

Author Contributions: Conceptualization, M.L. and M.D.; Data curation, M.D.; Formal analysis, M.L.; Methodology, J.K. and M.L.; Project administration, M.L.; Resources, M.L.; Supervision, M.D and M.L.; Validation, M.L. and J.K.; Writing—Original draft preparation, M.L.; Writing—review and editing J.K. and M.L. All authors have read and agreed to the published version of the manuscript.

Funding: This research received no external funding.

Institutional Review Board Statement: Not applicable.

Informed Consent Statement: Not applicable.

Data Availability Statement: No new data were created or analyzed in this study. Data sharing is not applicable to this article.

Acknowledgments: This work was supported by the Slovak Research and Development Agency under contract No. APVV-18-0526 and APVV-21-0027.

Conflicts of Interest: The authors declare no conflict of interest.

References

1. Sajjad, M.; Rasul, M.G. Review on the Existing and Developing Underground Coal Gasification Techniques in Abandoned Coal Seam Gas Blocks: Australia and Global Context. In Proceedings of the 1st International e—Conference on Energies, Virtual Event, 14–31 March 2014; pp. 1–16. Available online: http://sciforum.net/conference/ece-1 (accessed on 15 February 2022).

2. Bhutto, A.W.; Bazmi, A.A.; Zahedi, G. Underground coal gasification: From fundamentals to applications. *Prog. Energy Combust. Sci.* **2013**, *39*, 189–214. [CrossRef]
3. Couch, G. *Underground Coal Gasification*; Report No.: Contract No.: CCC/151; IEA Clean Coal Centre: London, UK, 2009.
4. Kačur, J.; Kostúr, K. *Underground Coal Gasification by Thermal Decomposition*; Research Report of Project APVV-0582-06; TU FBERG, ÚRaIVP: Kosice, Slovakia, 2008.
5. Laciak, M.; Durdán, M.; Kačur, J.; Flegner, P.; Benková, M. The Prediction of Possibilities of CO Poisoning and Explosion during Syngas Leakage in the UCG Process. *Processes* **2021**, *9*, 1912. [CrossRef]
6. Wałowski, G. Gas Permeability Model for Porous Materials from Underground Coal Gasification Technology. *Energies* **2021**, *14*, 4462. [CrossRef]
7. Pankiewicz-Sperka, M.; Kapusta, K.; Basa, W.; Stolecka, K. Characteristics of Water Contaminants from Underground Coal Gasification (UCG) Process—Effect of Coal Properties and Gasification Pressure. *Energies* **2021**, *14*, 6533. [CrossRef]
8. Skrzypkowski, K.; Zagórski, K.; Zagórska, A. Determination of the Extent of the Rock Destruction Zones around a Gasification Channel on the Basis of Strength Tests of Sandstone and Claystone Samples Heated at High Temperatures up to 1200 °C and Exposed to Water. *Energies* **2021**, *14*, 6464. [CrossRef]
9. Kapusta, K. Effect of Lignite Properties on Its Suitability for the Implementation of Underground Coal Gasification (UCG) in Selected Deposits. *Energies* **2021**, *14*, 5816. [CrossRef]
10. Bazaluk, O.; Lozynskyi, V.; Falshtynskyi, V.; Saik, P.; Dychkovskyi, R.; Cabana, E. Experimental Studies of the Effect of Design and Technological Solutions on the Intensification of an Underground Coal Gasification Process. *Energies* **2021**, *14*, 4369. [CrossRef]
11. Durdán, M.; Benková, M.; Laciak, M.; Kačur, J.; Flegner, P. Regression Models Utilization to the Underground Temperature Determination at Coal Energy Conversion. *Energies* **2021**, *14*, 5444. [CrossRef]
12. Kačur, J.; Laciak, M.; Durdán, M.; Flegner, P. Model-Free Control of UCG Based on Continual Optimization of Operating Variables: An Experimental Study. *Energies* **2021**, *14*, 4323. [CrossRef]
13. Zelenak, S.; Skvarekova, E.; Senova, A.; Wittenberger, G. The Usage of UCG Technology as Alternative to Reach Low-Carbon Energy. *Energies* **2021**, *14*, 3718. [CrossRef]

energies

Article

Characteristics of Water Contaminants from Underground Coal Gasification (UCG) Process—Effect of Coal Properties and Gasification Pressure

Magdalena Pankiewicz-Sperka [1,*], Krzysztof Kapusta [1], Wioleta Basa [1] and Katarzyna Stolecka [2]

1 Department of Energy Saving and Air Protection, Główny Instytut Górnictwa, Central Mining Institute, Plac Gwarków 1, 40-166 Katowice, Poland; kkapusta@gig.eu (K.K.); wbasa@gig.eu (W.B.)
2 Department of Power Engineering and Turbomachinery, Silesian University of Technology, Konarskiego 18, 44-100 Gliwice, Poland; katarzyna.stolecka@polsl.pl
* Correspondence: mpankiewicz@gig.eu; Tel.: +48-32-324-6536; Fax: +48-32-324-6522

Abstract: One of the most important issues during UCG process is wastewater production and treatment. Condensed gasification wastewater is contaminated by many hazardous compounds. The composition of the generated UCG-derived wastewater may vary depending on the type of gasified coal and conditions of the gasification process. The main purpose of this study was a qualitative and quantitative characterization of the UCG wastewater produced during four different UCG experiments. Experiments were conducted using semi-anthracite and bituminous coal samples at two distinct pressures, i.e., 20 and 40 bar. The conducted studies revealed significant relationships between the physicochemical composition of the wastewater and the coal properties as well as the gasification pressure. The strongest impact is noticeable in the case of organic pollutants, especially phenols, BTEX and PAH's. The most abundant group of pollutants were phenols. Conducted studies showed significantly higher concentration levels for bituminous coal: 29.25–49.5 mg/L whereas for semi-anthracite effluents these concentrations were in much lower range 2.1–29.7 mg/L. The opposite situation occurs for BTEX, higher concentrations were in wastewater from semi-anthracite gasification: 5483.1–1496.7 μg/L, while in samples from bituminous coal gasification average BTEX concentrations were: 2514.3–1354.4 μg/L. A similar relationship occurs for the PAH's concentrations. The higher values were in case of wastewater from semi-anthracite coal experiments and were in range 362–1658 μg/L while from bituminous coal gasification PAH's values are in lower ranges 407–1090 μg/L. The studies conducted have shown that concentrations of phenols, BTEX and PAH's decrease with increasing pressure. Pearson's correlation analysis was performed to enhance the interpretation of the obtained experimental data and showed a very strong relationship between three parameters: phenols, volatile phenols and COD_{cr}.

Keywords: underground coal gasification; SNG; UCG wastewater; environmental impact assessment; correlation analysis; effluents

Citation: Pankiewicz-Sperka, M.; Kapusta, K.; Basa, W.; Stolecka, K. Characteristics of Water Contaminants from Underground Coal Gasification (UCG) Process—Effect of Coal Properties and Gasification Pressure. *Energies* **2021**, *14*, 6533. https://doi.org/10.3390/en14206533

Academic Editor: Marek Laciak

Received: 4 August 2021
Accepted: 4 October 2021
Published: 12 October 2021

1. Introduction

Nowadays meeting the challenges of energy supply safety and provision of competitive energy costs is one of the most important challenges in the energy sector today. Despite the current ecological trends towards shifting to renewable energy and green resources, fossil fuels and coal will still be a major source of energy in a near future [1,2]. Coal has been and still is one of the most crucial primary energies and contributes approximately 65% of the total fossil fuel reserves in the world [3]. It is estimated that 45% of global energy demand will be covered by coal consumption by 2030 [2,4]. However, conventional coal mining has become more difficult and controversial. Ecological and economic factors stimulate searching for new ways and solutions for use of coal reserves. One of them is underground coal gasification (UCG) which offers many potential advantages over the traditional mining methods [5,6]. UCG is a method of in-situ (directly in the underground coal seam) thermochemical coal conversion into a synthetic gas [7–9]. The basis of the UCG process is direct injection of gasifying reagents to the ignited coal seam and receiving the gas product at the surface [10]. Compared to traditional mining UCG process has lower surface impact and hence may contribute to the reduction of air pollutants and greenhouse gas emission [11]. There are several process techniques for the UCG described in detail in the literature [11–14]. The final gas composition is mainly H_2, CH_4, CO and CO_2. The most desirable product for UCG process is methane, which strongly improve calorific value of gas [1,15]. Methane is formed in methanation reaction and directly from solid carbon in hydrogenation reaction [1]:

$$CO + 3H_2 \rightarrow CH_4 + H_2O \ (\Delta H = -206 \text{ kJ/mol})$$

$$C + 2H_2 \rightarrow CH_4 \ (\Delta H = -91 \text{ kJ/mol})$$

Methane rich gas called synthetic natural gas (SNG) can be used as a chemical feedstock or as a fuel for power generation [1,16]. SNG seems to be a future fuel and an essential component in the energy production, which will make several energy-intense industries more efficient and sustainable, while reducing their carbon footprint. However, every thermochemical coal processing technology is associated with environmental impact assessment. One of the most important issues is wastewater production and its treatment. The raw UCG product gas, apart from tar compounds and particulates (coal and ash) contains water vapour, mainly derived from the evaporation of coal moisture, the coal pyrolysis (pyrogenic water) or from hydrogen combustion. These gas components tend to condense onto the cooler parts of the facilities, such as the internal surfaces of gas pipelines or in the gas-treatment module particular devices (e.g., water scrubber). These condensed processing wastewater is contaminated by many hazardous compounds such as polycyclic aromatic hydrocarbons (PAHs), phenols, monoaromatic compounds including benzene, toluene, ethylbenzene and xylene [10,17–20]. Heavy metals are another group of UCG-derived contaminants [10,17,18]. Due to its specific nature, the UCG wastewater requires an appropriately tailored treatment technique. In 1988 Bryant et al. evaluate the biological treatability of wastewater from the UCG pilot installation in Hanna, Wyoming [21]. Zhang et al. propose pretreatment of wastewater generated during coal gasification by acidification demulsion [22]. A large number of toxic compounds present in UCG wastewater are difficult to decompose if only biological methods are used [23]. Thomas et al. presents the possibility of phenol removal from UCG effluents by using coagulation-flocculation and the H_2O_2/UV Process [24]. Treatment of coal gasification wastewater by catalytic oxidation with trace ozone is another promising technique [25]. In recent years there have been several new developments involving biological coupling processes to treat coal gasification wastewater. Biological coupling treatment methods including: conventional biological processes, the combination of adsorption and biotechnology processes, biological enhancement technologies, co-metabolism technologies and the combination of advanced oxidation and biotechnology [23–30]. The development of an appropriate treatment method to remove pollutants from UCG wastewater is of utmost importance for the successful implementation of this technology. However, the composition of the generated UCG-derived wastewater may vary depending on the type of gasified coal and conditions of the gasification process.

The main aim of the study was to conduct the qualitative and quantitative characterization of UCG wastewater generated during four different ex situ UCG experiments. The effluents were collected during the experiments in order to correlate the compositions and concentrations of produced contaminants with the coal properties (coal type) and gasification conditions.

2. Materials and Methods

2.1. Coal Samples and UCG Experiments

The four UCG experiments were carried out in an ex situ UCG installation located in the Clean Coal Technology Centre of the Central Mining Institute (Mikołów, Poland). The experimental installation enables simulation of the UCG process in surface conditions. The schematic view of the installation and wastewater sampling point are presented in Figures 1 and 2.

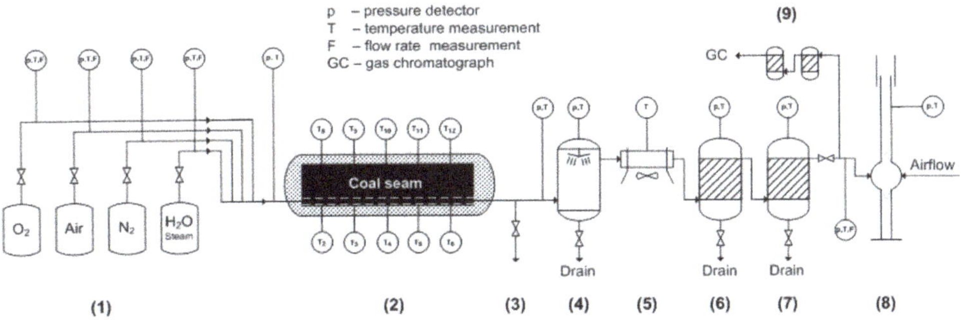

Figure 1. Schematic view of the ex-situ high pressure UCG installation. Reproduced from K. Kapusta et al. [1]. **(1)** reagent supply system, **(2)** gasification reactor, **(3)** tar sampling point, **(4)** water scrubber—wastewater sampling point, **(5)** air cooler for process gas, **(6,7)** gas separators, **(8)** thermal combustor, **(9)** gas purification module for GC analysis.

Figure 2. Water scrubber—wastewater sampling point.

Experiments were conducted using two different coal samples. Coal samples were gathered from two various locations. The first semi-anthracite "Six feet" coal was obtained from an open cast coal mine near Merthyr Tydfil (South Wales, UK) and the second one bituminous coal was obtained from the "Wesoła" coal mine located in Mysłowice (Upper Silesia, Poland). Detailed parameters of used coals are presented in Table 1. The raw coal samples were tested for 18 elements, including selected metals and metalloids being considered the most important for the aquatic environment. The results obtained are presented in Table 2.

Table 1. Characteristics of coals used for the UCG experiments.

Parameter	Coal	
	"Six-Feet" Semi-Anthracite	"Wesoła" Bituminous
	As received	
Total Moisture W_t^r, %	1.15 ± 0.40	3.60 ± 0.40
Ash A_t^r, %	4.61 ± 0.30	8.74 ± 40
Volatiles V^r, %	9.92 ± 0.12	27.67 ± 0.50
Total Sulphur S_t^r, %	1.55 ± 0.04	0.31 ± 0.02
Calorific value Q_i^r, kJ/kg	33,416 ± 220	28,798 ± 200
	Analytical	
Moisture W^a, %	0.84 ± 0.30	2.18 ± 0.27
Ash A^a, %	4.62 ± 0.30	8.87 ± 0.63
Volatiles V^a, %	9.95 ± 0.13	28.08 ± 0.92
Combustion Heat Q_s^a, kJ/kg	34,414 ± 228	30,317 ± 161
Calorific value Q_i^a, kJ/kg	33,527 ± 221	29,258 ± 201
Total Sulphur S^a, %	1.55 ±0.04	0.31 ± 0.08
Carbon C_t^a, %	87.31 ± 0.66	75.35 ± 1.13
Hydrogen H_t^a, %	3.97 ± 0.28	4.61 ±0.40
Nitrogen N^a, %	1.29 ± 0.12	1.20 ± 0.22
Oxygen O_d^a, %	0.50 ± 0.05	7.65 ± 0.1
Specific Gravity, g/cm^3	1.35 ± 0.028	1.40 ±0.018
Vitrinite reflectance, R_o, %	1.67 ± 0.03	0.91 ± 0.03
Vitrinite, V, vol.%	72 ± 6	59 ± 6
Liplinite, L, vol.%	0 ± 1	6 ± 4
Inertinite, I, vol.%	28 ± 3	35 ± 7
Mineral matter, MM, vol.%	2 ± 1	4 ± 3

Table 2. Concentrations of metals and metalloids in raw coals.

Element	"Six-Feet" Semi-Anthracite	"Wesoła" Bituminous
	mg/kg (ppm)	
As	10	0
B	14	18
Cd	0	1
Co	10	0.5
Cr	73	0.3
Cu	25	13
Hg	0.22	0.02
Mn	218	357
Mo	4	0.1
Ni	52	2.6
Pb	27	0.8
Sb	17	0.4
Se	0	2.2
Zn	14	8.1
	% mass	
Al	1.05	0.07
Fe	1.04	1.43
K	0.09	0.002
Ti	0.04	0.001

All gasification tests were conducted for a period of 96 h and under two distinct pressure regimes—20 and 40 bar. The general summary of the UCG experiments conducted is presented in Table 3.

Table 3. General summary of UCG experiments [1].

Coal Type	Semi-Anthracite "Six Feet" (South Wales, UK)	Semi-Anthracite "Six Feet" (South Wales, UK)	Bituminous "Wesoła" Coal (Upper Silesia, Poland)	Bituminous "Wesoła" Coal (Upper Silesia, Poland)
Gasification Reagent	O_2/H_2O	O_2/H_2O	O_2/H_2O	O_2/H_2O
Gasification Pressure, bar	20	40	20	40
Experiment duration	96	96	96	96
Average Gas Production Rate, Nm^3/h	9.0	9.4	9.3	9.4
Gas Yield, Nm^3/kg of coal consumed	1.98	1.98	1.77	1.70
Gas calorific value, Q, MJ/Nm^3	11.7	12.1	9.2	10.4
Coal gasified, kg	436.1	455.5	504.0	530.2
Total wastewater production, kg	46.5	38.6	67.3	55.2

To investigate the effect of coal type and gasification pressure the oxidant supply rates were the same in all experiments. During first 24 h of the process, oxygen was used as a gasifying agent, with constant flow5 Nm^3/h. After 24 h the processes were carried out with oxygen and water with flow ratio 5 Nm^3/h and 2.5 kg/h respectively.

2.2. Post-Processing Water Sampling

The UCG effluents produced in water scrubber were collected after completion of each gasification experiment. They represented the average sample of wastewater for given gasification experiment. After sampling, the wastewater were transported to the laboratory for chemical analyses. Coal tars and other undissolved residues were removed by vacuum filtration WhatmanTM Glass Microfiber Filters GF/CTM (GE Healthcare UK Limited, Hatfield, UK), and filtrates were subsequently stored at 4 °C until analysed.

2.3. Chemical Analyses

The chemical analyses were carried out according to standard analytical methods. The conductivity, pH and COD_{Cr} (chemical oxygen demand) were determined as typical nonspecific industrial wastewater parameters. Following inorganic parameters were also determined: total ammonia nitrogen, chlorides, cyanides, sulphates, sulphides and 17 metal and metalloid trace elements (Mn, Fe, Sb, As, B, Cr, Zn, Al, Cd, Co, Cu, Mo, Ni, Pb, Hg, Se, Ti). Organic analysis included benzene with its three alkyl homologues: toluene, ethylbenzene and xylene (BTEX), total phenols and 15 polycyclic aromatic hydrocarbons (PAHs). To determine pH and conductivity potentiometry and conductometry methods were used according to PN-EN ISO 10523: 2012 and PN-EN 27888:1999 standards. COD_{Cr} index was determined by spectrophotometric method according to PN-ISO 15705: 2005. Ammonia nitrogen was determined by Flow Injection Analysis (FIA) with gaseous diffusion and spectrophotometric detection according to PN-EN ISO 11732: 2007). The chlorides were determined according to PN-ISO 9297: 1994. The cyanides and the volatile phenols were determined by segment flow analysis (SFA) with spectrophotometric detection according to PN-EN ISO 14403-2:2012 and PN-EN ISO 14402:2004. Sulphates were determined according to PN-ISO 9280: 2002. Flow Injection Analysis (FIA) with spectrophotometric detection was used to determined sulphides. To determined metals and metalloid trace elements inductively coupled plasma-optical emission spectroscopy (ICP-OES) was used (PN-EN ISO 11885: 2009). For the BTEX and phenols analysis the Agilent Technologies 7890A chromatograph coupled with a static headspace auto sampler Agilent 7697A and FID detector was applied. The chromatographic column was DB-5MS (30 m, 0.25 mm, 0.5 µm). For determination of PAHs high-performance liquid chromatography was applied using Agilent Technologies HPLC Series chromatograph equipped with fluorescence detector on Agilent ZORBAX Eclipse PAH column (3.0 mm × 250 mm, 5 µm).

2.4. Linear Correlation Analysis

Pearson's correlation analysis was performed to enhance the interpretation of the obtained experimental data. It is known as a valuable method of measuring the association between variables data because it is based on the method of covariance. Pearson's correlation analysis gives information about the magnitude of the correlation and direction of the relationship. The values of the Pearson coefficient "r" can fluctuate from −1 to 1. An $r = -1$ indicates a perfect negative linear relationship, an $r = 0$ indicates no linear relationship, and an $r = 1$ indicates a perfect positive linear relationship between variables. The closer the indicator is to 1, the greater the correlation occurs. In statistical analysis, it is assumed that the values >0.7 indicating significant correlation between the variables. Input data were physicochemical parameters of obtained wastewater samples from all four UCG experiments.

3. Results and Discussion

The average physicochemical characteristics of the post processing water samples obtained during all four UCG experiments are presented in the Table 4. Conducted study revealed significant differences in the qualitative and quantitative characteristics of the tested water samples. The differences obtained were related to both the type of the coal used and the applied gasification pressure. The results of the Pearson's correlation analysis are presented in Table 5. The values of the Pearson coefficient >0.7 are bolded.

Table 4. Average values of physicochemical parameters determined in the UCG effluents from semi-anthracite and bituminous coal experiments.

Parameters	Unit	Semi-Anthracite Coal		Bituminous Coal	
		20 Bar	40 Bar	20 Bar	40 Bar
pH	pH	6.4	5.2	5.3	4.9
Conductivity	µS/cm	1228.38	253.38	942	1006.71
COD_{Cr}	mg/L O_2	151.63	48.63	322.71	185.91
Ammonia nitrogen	mg/L N	160.11	11.68	96.41	95.74
Chlorides	mg/L	11.15	11.68	29.18	45.94
Cyanides	mg/L	1.11	1.43	1.7	0.87
Total phenols volatile	mg/L	8.45	0.87	17.04	24.46
Sulphates	mg/L	33.51	47.66	42.86	52.97
Sulphides	mg/L	1.04	0.04	0.97	0.02
Mn	mg/L	0.017	0.021	0.018	0.012
Fe	mg/L	0.823	0.284	0.131	0.245
Sb	mg/L	0.036	0.121	0.064	0.013
As	mg/L	0.036	<0.02	<0.01	<0.01
B	mg/L	0.072	0.056	0.130	0.252
Cr	mg/L	0.013	0.012	0.010	0.006
Zn	mg/L	0.021	0.499	0.320	0.200
Al	mg/L	0.031	0.046	0.029	0.023
Cd	mg/L	<0.0005	0.001	<0.0005	<0.0005
Co	mg/L	0.004	0.003	<0.003	<0.003
Cu	mg/L	0.005	0.010	0.009	0.002
Mo	mg/L	0.005	<0.005	0.026	<0.005
Ni	mg/L	0.098	0.312	0.051	0.027
Pb	mg/L	<0.005	0.064	0.046	0.060
Hg	mg/L	<0.0005	<0.0005	<0.0005	<0.0005
Se	mg/L	0.016	0.017	0.036	0.027
Ti	mg/L	<0.0005	0.001	0.001	<0.0005
Total BTEX	µg/L	5483.13	1496.73	2514.32	1354.37
Including benzene	µg/L	4156.08	1341.43	2196.75	1059.07
Total PAH	µg/L	1657.98	361.99	1090.34	407.2
Including Naphthalene	µg/L	1321.25	320.88	905	305.74
Total Phenols	mg/L	29.73	2.14	49.46	29.25

Table 5. Pearson correlation matrix—results of the linear correlation analysis of physicochemical parameters of UCG wastewater.

	pH	Cond.	COD_{Cr}	NH_4^+	Cl^-	CN^-	Volatile Phenols	SO_4^{2-}	S^{2-}	Fe	B	Zn	Al	Ni	Pb	Se	BTEX	PAH	Phenols
pH	1.00																		
Cond.	0.55	1.00																	
COD_{Cr}	0.20	0.56	1.00																
NH_4^+	0.63	0.99	0.53	1.00															
Cl^-	−0.57	−0.07	0.23	−0.16	1.00														
CN^-	0.19	0.00	0.19	0.02	−0.11	1.00													
Volatile phenols	0.39	0.77	0.87	0.75	0.07	0.18	1.00												
SO_4^{2-}	−0.64	0.05	0.10	−0.02	0.33	−0.14	0.04	1.00											
S^{2-}	0.31	0.44	0.13	0.46	−0.14	0.02	0.15	0.13	1.00										
Fe	−0.10	−0.15	−0.14	−0.16	−0.03	0.04	−0.15	0.04	0.06	1.00									
B	−0.16	0.59	0.57	0.52	0.41	−0.25	0.65	0.40	−0.07	−0.13	1.00								
Zn	−0.57	−0.55	−0.44	−0.56	−0.02	−0.02	−0.46	0.30	−0.34	−0.20	−0.19	1.00							
Al	−0.49	−0.18	−0.13	−0.23	0.28	0.24	−0.10	0.59	−0.21	−0.06	0.02	0.53	1.00						
Ni	−0.26	−0.28	−0.35	−0.27	−0.17	0.23	−0.28	0.34	−0.07	0.45	−0.24	0.50	0.66	1.00					
Pb	−0.71	−0.30	−0.15	−0.35	0.36	−0.33	−0.22	0.60	−0.21	−0.04	0.22	0.58	0.58	0.38	1.00				
Se	0.35	0.87	0.61	0.83	0.09	0.03	0.83	0.14	0.25	−0.10	0.66	−0.41	−0.11	−0.25	−0.14	1.00			
BTEX	0.35	0.21	0.18	0.22	−0.16	0.35	0.21	−0.20	0.45	0.66	−0.21	−0.50	−0.21	0.16	−0.44	0.16	1.00		
PAH	0.37	0.63	0.31	0.64	−0.07	−0.12	0.32	0.15	0.89	−0.09	0.16	−0.50	−0.25	−0.28	−0.24	0.37	0.35	1.00	
Phenols	0.43	0.67	0.75	0.66	−0.09	0.31	0.83	−0.06	0.34	−0.08	0.38	−0.53	−0.18	−0.32	−0.34	0.63	0.39	0.50	1.00

3.1. Coal Type Effect

As can be seen from the Table 4 all analysed water samples exhibit high values of the COD_{Cr} parameter, which is typical for effluents from the thermochemical processing of coal. The much higher COD_{Cr} values were observed in water samples from gasification of bituminous coal, ranged from 185.9 mg/L_{O2} to 322.7 mg/L_{O2}, while for semi-anthracite coal this parameter was in the range from 48.6 mg/L_{O2} to 151.6 mg/L_{O2}. pH of analysed water samples was slightly higher for semi-anthracite experiments, fluctuating within 5.2–6.4 level and 4.9–5.3 for bituminous coal. Ammonia nitrogen levels for bituminous coal remained relatively constant from 95.7 mg/L to 96.4 mg/L, while for semi-anthracite coal wastewater there was a wide concentration range from 11.7 mg/L to 160.1 mg/L. This situation is determined by pH values, which were in a wider range and fluctuated more during the gasification of semi-anthracite coal. For chlorides there was the opposite situation and in effluents from gasification of semi-anthracite coal concentrations were in the lower range 11.2–11.7 mg/L while for bituminous coal wastewater levels were higher and fluctuated in a wider range from 29.2 mg/L to 45.9 mg/L. In all wastewater samples low concentration levels of cyanides and sulphides were observed. Sulphates levels were relatively higher for wastewater from bituminous coal gasification and were from 42.9 mg/L to 53.0 mg/L while for semi-anthracite coal concentration values were in range of 33.5–47.7 mg/L. The conducted studies have shown concentrations of metals and metalloids in all studied water samples were at very low levels (Table 4). Among the 17 of metals and metalloids, 9 of them (Mn, As, Cr, Cd, Co, Cu, Mo, Hg and Ti) were identified in concentrations below the lower detection limit or in an amount not exceeding 0.036 mg/L (for As). For the rest metals and metalloids concentrations were above lower detection limits, but still at very low levels. While in raw coals the highest values were for Mn and were 218 mg/kg and 357 mg/kg for semi-anthracite and bituminous coal respectively, the highest values in effluents were observed for Fe. For semi-anthracite wastewater concentrations varied from 0.284 to 0.823 mg/L, while for bituminous effluents Fe levels were lower and range from 0.131 mg/L to 0.245 mg/L. Concentrations of metals and metalloids occurring in the raw coal do not directly affect the composition of the wastewater generated during the UCG process. This is due to the fact that their concentrations are dependent on the solubility of the individual elements, which varies with pH and the presence of other compounds (background) in the sample. The wastewater which are formed during the process is water coming from condensation onto the cooler parts of the installations (e.g., in particular devices of the gas-treatment module). Composition of obtained wastewaters is therefore mainly determined by organic contaminants originating from the tars which are generated during the gasification process. Therefore, the studies carried out confirmed that type of coal used for gasification experiments has a significant impact on concentration levels of organic compounds. Among all pollutants, organic compounds (phenols, BTEX, PAH) constituted the most significant group of contaminants in UCG wastewater samples. Comparison of selected organic contaminants concentrations in the wastewater from gasification experiments are presented in Figure 3.

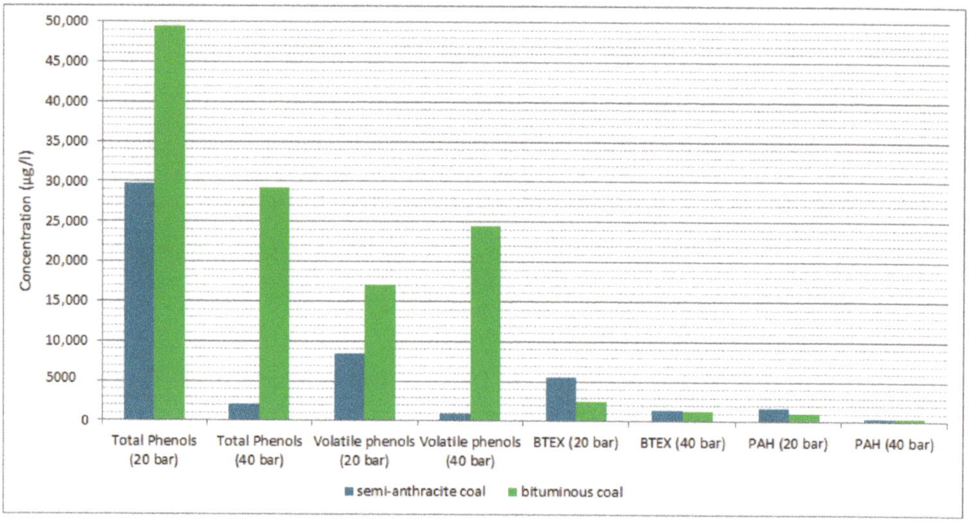

Figure 3. Concentrations of selected organic contaminants in wastewater from gasification experiments—coal type impact.

As can be seen from the Figure 3 due to high water affinity, the most abundant group of pollutants in analysed water samples were phenols. Conducted studies showed significantly higher concentration levels for bituminous coal wastewater, with values from 29.25 to 49.5 mg/L whereas for semi-anthracite effluents these concentrations were in much lower range 2.1–29.7 mg/L. An analogous situation exists for volatile phenols, where average concentrations in bituminous coal wastewater were 17 and 24 mg/L at 20 and 40 bar, while for semi-anthracite coal wastewater the average concentrations were proportionally lower 8.45 and 0.87 at 20 and 40 bar respectively. However the opposite situation occurs for BTEX levels. The conducted studies showed that higher concentrations occurs in wastewater from semi-anthracite gasification. BTEX average amounts are 5483.1 µg/L for 20 bar experiment and 1496.7 µg/L for 40 bar experiment, while in samples from bituminous coal gasification average BTEX concentrations were in lower range 2514.3–1354.4 µg/L. A similar relationship can be found for the PAH's concentrations. The higher values 362–1658 µg/L occurs in case of wastewater from semi-anthracite coal experiments. For wastewater from bituminous coal gasification PAH's values are in lower ranges 407–1090 µg/L.

3.2. *Effect of Gasification Pressure*

The conducted studies revealed some dependencies between coal gasification pressure and physicochemical composition of analysed post-processing water samples. It was observed that pressure affects such parameters as chloride and sulphate concentrations. As can be seen from Table 4 chloride release increases along with increasing pressure, especially for bituminous coal effluents, where chlorides levels were 29.18 mg/L and 45.94 mg/L for 20 and 40 bar respectively. The same situation occurs for sulphates concentrations. For 20 bar pressure sulphates levels were 33.5 mg/L for bituminous coal and 42.9 mg/L for semi-anthracite coal effluents. When process pressure increased to 40 bar, concentrations were also higher and were 47.7 mg/L and 53.0 mg/L for bituminous and semi-anthracite coal respectively.

Just as it was in the case of coal impact the impact of pressure is especially noticeable in the case of organic compounds such as phenols, BTEX and PAH. Comparison of selected wastewater organic contaminants from gasification of semi-anthracite and bituminous coal are presented in Figure 4.

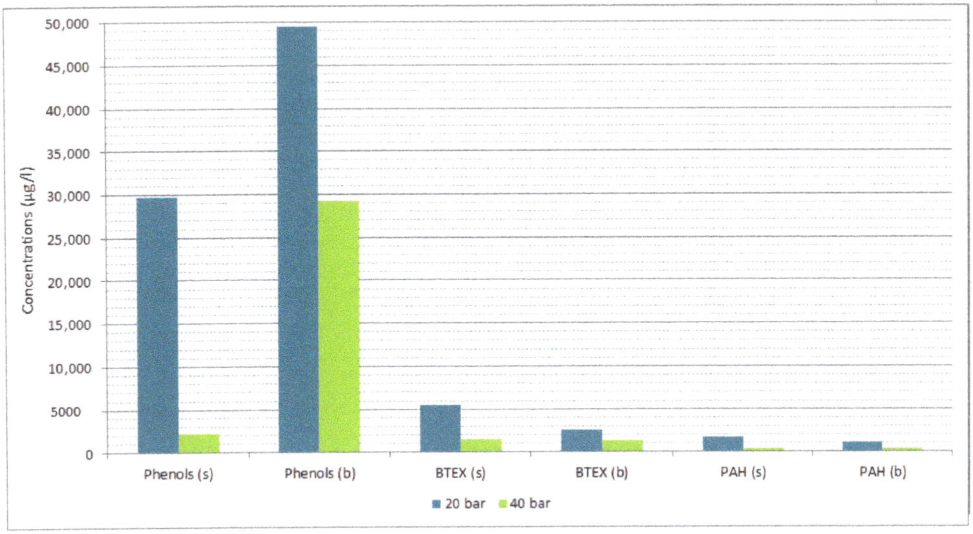

Figure 4. Comparison of selected wastewater organic contaminants from gasification of semi-anthracite and bituminous coal—pressure impact ((s)—semi-anthracite coal; (b)—bituminous coal).

The studies conducted have shown that concentrations of phenols decrease with increasing pressure. When gasification pressure was lower (20 bar) phenols concentrations were in the field of 29.7 mg/L and 49.46 mg/L for semi-anthracite and bituminous coal respectively. Whereas in the case of the high-pressure experiments, there was more than 10-fold decrease in phenols concentration for hard coal and almost halved decrease for bituminous coal, reaching values 2.14 mg/L and 29.25 mg/L respectively. This significant decrease in the concentration of phenols with the increase in gasification pressure resulted in a significant decrease in the value of COD_{Cr} parameter, which is strongly correlated with the concentration of phenols (Table 5). The same situation occurred with the BTEX values and with increasing pressure there were large decreases in BTEX concentrations. In the case of 20 bar hard coal gasification process, the average BTEX values in the studied effluents were 5483.1 μg/L, while for the high-pressure 40 bar process these values decreased more than threefold to 1496.7 μg/L. For effluents from bituminous coal gasification, the decrease was slightly lower, with BTEX values of 2514.2 μg/L at 20 bar and 1354.4 μg/L at 40 bar, respectively. The effect of pressure was also observed for PAH levels. As the pressure increases, there is a large decrease in PAH concentration in the studied wastewater samples from all four experiments. In the case of semi-anthracite coal experiment there is a decrease from 1658.0 μg/L to 362 μg/L. In the case of bituminous coal the difference is also significant, for the 20 bar experiment the PAH value was 1090.3 μg/L while for the 40 bar experiment the average value was 407.2 μg/L. For all discussed organic compounds groups the same dependence occurs, with the increase of pressure their concentration in the studied effluents decreases. It can be explained by volatility of these compounds. At lower pressure more of them are dissolved in the water phase. However, as the pressure increases, a greater release of the compounds into UCG gas takes place.

3.3. Pearson's Correlation Analysis

Correlation analysis (Table 5) showed a strong relationship between the conductivity of the studied effluents and the level of ammonia nitrogen. The Pearson's correlation coefficient was 0.99 which indicates an almost linear relationship between these two parameters for all four gasification experiments. Furthermore, correlation analysis showed a very strong relationship between three parameters: phenols, volatile phenols and COD_{cr}. The correlation coefficients were 0.87 and 0.75 for COD_{Cr}—phenols volatile and COD_{Cr}—total phenols respectively. On the other hand, correlation analysis showed no significant dependence between COD_{Cr} parameter and other toxic organic compounds concentrations such as BTEX or PAH. Although high toxicity of these compounds, the general toxicity of gasification wastewater is mainly determined by concentration of phenols [17]. The main reason for this may be the levels of BTEX and PAH concentrations, which are several times and in some cases even several dozen times lower than the levels of phenols. For metals and metalloids no significant correlations were observed. This can be explained by the low concentrations levels in studied wastewater samples. Only for Se correlation analysis showed a high correlation coefficient between Se and conductivity ($r = 0.87$), Se and NH_4^+ ($r = 0.83$) and Se—volatile phenols ($r = 0.83$).

4. Conclusions

The studies conducted revealed that the type of coal used and gasification pressure have a significant impact on the wastewater parameters. The conducted studies on the gasification effluents revealed significant relationships between the physicochemical composition of the wastewater and the coal properties as well as the gasification pressure. Regarding the impact of the used coal, influence on parameters such as pH and chloride can be observed. The pH of the obtained water samples was slightly higher for the semi-anthracite coal, whereas chloride levels were higher for effluents from gasification of bituminous coal. The water samples from bituminous coal gasification showed significantly higher levels of COD parameter. The studied water samples were characterised by a high concentration of organic compounds, therefore the strongest impact is noticeable in the case of these pollutants, especially volatile phenols, phenols, BTEX and PAH. Concentrations of volatile phenols and phenols were much higher for bituminous coal. However, for the BTEX and PAH levels, the opposite situation was observed and higher concentrations were in the case of wastewater from gasification of semi-anthracite coal. Gasification pressure has also noticeable impact on the composition of obtained gasification wastewater. As can be seen from the presented data, there is a greater release of chlorides along with increasing pressure, especially in the case of bituminous coal. The same situation also occurs for sulphates concentrations. As well as for the impact of the coal type, gasification pressure impact is the most significant in the case of organic compounds. As has been shown, their concentrations are inversely proportional to the gasification pressure. The conducted analysis showed that among the three main groups of organic pollutants: phenols, BTEX and PAHs, phenols were present at the highest concentrations. Therefore, it can be assumed that phenolic compounds will have the greatest impact on the toxicity level of the tested UCG wastewater. Correlation analysis showed also a strong relationship between the conductivity of the studied water samples and the level of ammonia nitrogen. The Pearson's correlation coefficient for these two parameters was 0.99 which indicates an almost linear relationship between them. The conducted research has shown that the composition of mineral matter of raw coals does not directly affect the composition of the UCG wastewater. This is because the concentrations of metals and metalloids are strongly pH dependent. Therefore, the composition of the obtained wastewater is determined mainly by organic pollutants derived from tars, which are generated in the gasification process. The conducted research has shown that UCG wastewater contains many hazardous pollutants and requires the selection of an appropriate treatment method, for example, such as for coking wastewater. The presented results can help in the development of an appropriate UCG wastewater treatment strategy depending on the coal used and gasification parameters.

Author Contributions: M.P.-S.: Conceptualization, Writing—original draft; K.K.: Conceptualization, Methodology, Project Administration, Funding Acquisition; W.B.: Experiments, Visualization; K.S.: Writing—review and editing, Visualization. All authors have read and agreed to the published version of the manuscript.

Funding: This work was a part of the MEGAPlus project supported by the EU Research Fund for Coal Steel, under the Grant Agreement number 800774—MEGAPlus—RFCS-2017 and Polish Ministry of Science and Higher Education under Grant Agreement No. 3996/FBWiS/2018/2.

Institutional Review Board Statement: Not applicable.

Informed Consent Statement: Not applicable.

Data Availability Statement: Supplementary data associated with this article can be found in the online version at https://doi.org/10.3390/en13061334 (accessed on 10 October 2021).

Conflicts of Interest: The authors declare no conflict of interest.

References

1. Kapusta, K.; Wiatowski, M.; Stańczyk, K.; Zagorscak, R.; Thomas, H.R. Large-scale experimental investigations to evaluate the feasibility of producing methane-rich gas (SNG) through under-ground coal gasification process. Effect of coal rank and gasification pressure. *Energies* **2020**, *13*, 1334. [CrossRef]
2. Knapp, R. Environmental challenges for the global coal industry. World Coal Institute, Coal/ Power Focus. *Asian J. Min.* **2000**, *208*, 15–18.
3. Chen, L.; Li, C.; Xu, B.; Xing, B.; Yi, G.; Huang, G.; Zhang, C.; Liu, J. Microbial degradation of organic pollutants in groundwater related to underground coal gasification. *Energy Sci. Eng.* **2019**, *7*, 2098–2111. [CrossRef]
4. Balat, M. Status of Fossil Energy Resources: A Global Perspective. *Energy Sources Part B Econ. Plan. Policy* **2007**, *2*, 31–47. [CrossRef]
5. Kapusta, K.; Stańczyk, K. Pollution of water during underground coal gasification of hard coal and lignite. *Fuel* **2011**, *90*, 1927–1934. [CrossRef]
6. Xu, M.; Xin, L.; Liu, W.; Hu, X.; Cheng, W.; Li, C.; Wang, Z. Study on the physical properties of coal pyrolysis in underground coal gasification channel. *Powder Technol.* **2020**, *376*, 573–592. [CrossRef]
7. Wiatowski, M.; Kapusta, K. Evolution of tar compounds in raw gas from a pilot-scale underground coal gasification (UCG) trial at Wieczorek mine in Poland. *Fuel* **2020**, *276*, 118070. [CrossRef]
8. Bhutto, A.; Bazmi, A.; Zahedi, G. Underground coal gasification: From fundamentals to applications. *Prog. Energy Combust. Sci.* **2013**, *39*, 189–214. [CrossRef]
9. Khadse, A.; Qayyumi, M.; Mahajani, S.; Aghalayam, P. Underground coal gasification: A new clean coal utilization technique for India. *Energy* **2007**, *32*, 2061–2071. [CrossRef]
10. Pankiewicz-Sperka, M.; Stańczyk, K.; Płaza, G.; Kwaśniewska, J.; Nałęcz-Jawecki, G. Assessment of the chemical, microbiological and toxicological aspects of post-processing water from under-ground coal gasification. *Ecotoxicol. Environ. Saf.* **2014**, *108*, 294–301. [CrossRef]
11. Yang, D.; Koukouzas, N.; Green, M.; Sheng, Y. Recent development on underground coal gasifica-tion and subsequent CO_2 storage. *J. Energy Inst.* **2016**, *89*, 469–484. [CrossRef]
12. Perkins, G. Underground coal gasification. Part I: Field demonstrations and proces performance. *Prog. Energy Combust. Sci.* **2018**, *67*, 158–187. [CrossRef]
13. Wiatowski, M.; Kapusta, K.; Świądrowski, J.; Cybulski, K.; Ludwik-Pardała, M.; Grabowski, J.; Stańczyk, K. Technological aspects of underground coal gasification in the Experimental "Barbara" Mine. *Fuel* **2015**, *159*, 454–462. [CrossRef]
14. Shafirovich, E.; Varma, A. Underground Coal Gasification: A Brief Review of Current Status. *Ind. Eng. Chem. Res.* **2009**, *48*, 7865–7875. [CrossRef]
15. Kopyscinski, J.; Schildhauer, T.J.; Biollaz, S.M. Production of synthetic natural gas (SNG) from coal and dry biomass–A technology review from 1950 to 2009. *Fuel* **2010**, *89*, 1763–1783. [CrossRef]
16. van der Meijden, C.M.; Veringa, H.J.; Rabou, L.P. The production of synthetic natural gas (SNG): A comparison of three wood gasification systems for energy balance and overall efficiency. *Biomass-Bioenergy* **2010**, *34*, 302–311. [CrossRef]
17. Kapusta, K.; Stańczyk, K. Chemical and toxicological evaluation of underground coal gasification (UCG) effluents. The coal rank effect. *Ecotoxicol. Environ. Saf.* **2015**, *112*, 105–113. [CrossRef] [PubMed]
18. Strugała -Wilczek, A.; Basa, W.; Kapusta, K.; Soukup, K. In situ sorption phenomena can mitigate potential negative environmental effects of underground coal gasification (UCG)—An experimental study of phenol removal on UCG-derived residues in the aspect of contaminant retardation. *Eco-Toxicol. Environ. Saf.* **2021**, *208*, 111710. [CrossRef] [PubMed]
19. Grabowski, J.; Korczak, K.; Tokarz, A. Aquatic risk assessment based on the results of research on mine waters as a part of a pilot underground coal gasification process. *Process. Saf. Environ. Prot.* **2020**, *148*, 548–558. [CrossRef]

20. Smoliński, A.; Stańczyk, K.; Kapusta, K.; Howaniec, N. Chemometric study of the ex situ under-ground coal gasification wastewater experimental data. *Water Air Soil Pollut.* **2012**, *223*, 5745–5758. [CrossRef]
21. Bryant, C.W.; Cawein, C.C.; King, P.H. Biological Treatability of In Situ Coal Gasification Wastewater. *J. Environ. Eng.* **1988**, *114*, 400–414. [CrossRef]
22. Zhang, W.; Ma, J.; Yang, S.; Zhang, T.; Li, Y. Pretreatment of Coal Gasification Wastewater by Acidification Demulsion. *Chin. J. Chem. Eng.* **2006**, *14*, 398–401. [CrossRef]
23. Shi, J.; Han, Y.; Xu, C.; Han, H. Biological coupling process for treatment of toxic and refractory compounds in coal gasification wastewater. *Rev. Environ. Sci. Bio/Technol.* **2018**, *17*, 765–790. [CrossRef]
24. Thomas, M.; Zdebik, D.; Niewiara, E. Removing Phenols from Post-Processing Wastewater Originating from Underground Coal Gasification Using Coagulation-Flocculation and the H2O2/UV Process. *Pol. J. Environ. Stud.* **2018**, *27*, 2757–2763. [CrossRef]
25. Zhang, G.; Wan, X.; Li, W.; Yang, Y.; Wang, L. Advanced treatment of coal gasification wastewater by catalytic oxidation with trace ozone. *Chin. J. Environ. Eng.* **2013**, *7*, 263–267.
26. Oller, I.; Malato, S.; Sánchez-Pérez, J.A. Combination of Advanced Oxidation Processes and bio-logical treatments for wastewater decontamination—A review. *Sci. Total. Environ.* **2011**, *409*, 4141–4166. [CrossRef]
27. Li, P.; Ailijiang, N.; Cao, X.; Lei, T.; Liang, P.; Zhang, X.; Huang, X.; Teng, J. Pretreatment of coal gasification wastewater by adsorption using activated carbons and activated coke. *Colloids Surf. A: Physicochem. Eng. Asp.* **2015**, *482*, 177–183. [CrossRef]
28. Priyadharshini, S.; Bakthavatsalam, A. A comparative study on growth and degradation behavior of C. pyrenoidosa on synthetic phenol and phenolic wastewater of a coal gasification plant. *J. Environ. Chem. Eng.* **2019**, *7*, 103079. [CrossRef]
29. Oberoi, A.S.; Philip, L. Performance evaluation of attached biofilm reactors for the treatment of wastewater contaminated with aromatic hydrocarbons and phenolic compounds. *J. Environ. Chem. Eng.* **2017**, *5*, 3852–3864. [CrossRef]
30. Wang, W.; Han, H.; Yuan, M.; Li, H.; Fang, F.; Wang, K. Treatment of coal gasification wastewater by a two-continuous UASB system with step-feed for COD and phenols removal. *Bioresour. Technol.* **2011**, *102*, 5454–5460. [CrossRef] [PubMed]

energies

MDPI

Article

The Usage of UCG Technology as Alternative to Reach Low-Carbon Energy

Stefan Zelenak [1], Erika Skvarekova [2,*], Andrea Senova [2] and Gabriel Wittenberger [2]

1 Hornonitrianske Bane Prievidza, a.s., Matice Slovenskej 10, 971 01 Prievidza, Slovakia; szelenak60@gmail.com
2 Faculty of Mining, Ecology, Process Control and Geotechnologies, Institute of Earth's Resources, Technical University of Kosice, Park Komenskeho 19, 040 01 Kosice, Slovakia; andrea.senova@tuke.sk (A.S.); gabriel.wittenberger@tuke.sk (G.W.)
* Correspondence: erika.skvarekova@tuke.sk; Tel.: +421-55-6023148

Abstract: Countries of the European Union have stated transition to carbon-neutral economy until the year of 2050. Countries with a higher share of coal-fired power generation currently have no solution to end their combustion and use clean, emission-free energy immediately. The solution to this problem in the energy industry appears to be the increased use of natural gas, which significantly reduces CO_2 emissions. In this article, we investigated the possibility of using coal in situ, using UCG (underground coal gasification) technology. We focused on verified geological, hydrogeological, and tectonic information about the selected brown coal deposit in Slovakia. This information has been assessed in research projects in recent years at the Technical University. From the abovementioned information, possible adverse factors were evaluated. These factors affect the rock environment around the underground generator by UCG activity. As part of the process management, measures were proposed to eliminate the occurrence of pollution and adverse effects on the environment. In the final phase of the UCG technology, we proposed to carry out, in the boreholes and in the generator cavity, water flushing and subsequent grouting. The proposed are suitable materials for solidification and stabilization. Results of this article´s solutions are crucial in the case of usage of this so-called clean technology, not only in Slovakia but also worldwide.

Keywords: low-carbon energy; UCG technology; grouting; solidification soil; soil air; statistic model; soil contamination; atmospheric geochemical survey; environmental burden

Citation: Zelenak, S.; Skvarekova, E.; Senova, A.; Wittenberger, G. The Usage of UCG Technology as Alternative to Reach Low-Carbon Energy. *Energies* **2021**, *14*, 3718. https://doi.org/10.3390/en14133718

Academic Editor: Praveen Cheekatamarla

Received: 18 May 2021
Accepted: 12 June 2021
Published: 22 June 2021

1. Introduction

The European Union has set itself an ambitious target for the transition to a carbon-neutral economy. In 2050, some countries are expected to be carbon-neutral without any problems. However, in time, Europe will have to acknowledge that each country has started from a different position, with a different background in history. Energy is a sector where investment is made for decades to come. It is assumed that there will also be countries, such as Poland, which currently obtain more than 80 percent of their energy from coal. It is unrealistic to change the structure from full coal to carbon-free energy in 30 years, in addition to the impending social consequences. Moreover, the current disparities between the energies of the EU's western and eastern countries could create tensions in certain critical situations, and also determine the future and integrity of the European Union (EU).

The direction of European energy policy is determined mainly by the German economy. The Czech Republic and Slovakia have to adapt to the fact that the time for national self-sufficient energy has ended. Related to this is the transition to carbon-free technologies, although it is not yet clear what energy will look like after 2050. If Slovakia manages to put into operation the nuclear units in Mochovce, it should not have problems with the transition to emission-free technologies in the production of electricity. The Czech Republic and Poland may be in a more difficult position, in which the share of coal in production is more significant. The problem is also expected in the eastern part of Germany when switching to a carbon-neutral economy.

However, the situation in Slovakia has changed in recent years. In 2010, the Government of the Slovak Republic passed Resolution No. 47/2010, approved under the General Economic Interest, directing the volumes of production and supply of electricity and heat from domestic coal [1].

At the same time, this support also had a significant social dimension, which consisted in maintaining employment in the Horna Nitra and Zahorie region. This measure ensured the optimal level of coal mining and higher security of electricity supply as well as lower energy dependence of the Slovak Republic for the period until 2020 and prospectively until 2035. The largest consumer of coal was Slovenské elektrárne, a.s.—Novaky Power Plant (ENO) [2].

The security of energy supplies is important, of course, all in the context of international obligations, in particular, those relating to climate change and the environment, especially air protection [3].

The change occurred in 2018 with the adoption of the Government Resolution No. 580/2018, when there was a fundamental change: the end of support for the production of electricity from domestic coal in 2023. In practice, this means the premature closure of mining operations 7 years earlier than the original concept [4].

The government has already set targets where, after 2020, it will gradually replace classical extraction methods by underground coal gasification, thereby ensuring synthesis gas for electricity and heat or, more precisely, production for chemical use [5].

Measures have been set to achieve the objectives of:

- executing in situ research on underground coal gasification (2015);
- regularly assessing, in cooperation with the regulatory authority, the costs and benefits resulting from supporting electricity generation, optimizing costs, and increasing the efficiency of its production; and
- maintaining the general economic interest for the production and supply of electricity produced from domestic coal during optimization of electricity generation, as well as ensuring, through an appropriate regulatory framework, the return of investments necessary for fulfillment with the obligations under Directive 2010/75/EU on industrial emissions [5,6].

In order to achieve carbon-free energy, it is primarily necessary, traditional, and non-renewable sources such as coal, natural gas, and oil that need to be replaced by alternative energy sources such as photovoltaic, wind, and nuclear energy [7,8].

The abovementioned situation in the developing low-carbon energy forces Slovakia and other countries to think about the use of other technologies, in an effort to ensure the required conditions. The key question to be answered in order to make a significant contribution to environmental protection is: Which types of fuels would be the best to reduce nitrogen oxides' emissions and, at the same time, avoid the growth of other gaseous pollutants?

The answer may be: Natural gas and propane-butane both serve as widely available gaseous fuels. This would not create problems regarding the size and distribution of pollutant particles and many others that need to be taken into account when testing solid fuels [9].

One of the possibilities could be underground coal gasification in situ. The article was designed in such a way to draw attention to important properties of the deposit, which are closely related to the possibility of using UCG technology in the conditions of Slovakia.

The main advantages of the technology mentioned include reduction of two major greenhouse gases, namely, methane and carbon monoxide. All the technologies for capturing of methane are appreciated these days, as methane is emission gas that has up to 20 times greater effect in the atmosphere than CO_2. In the case of UCG technology, no emission gas is consumed, but the idea of bringing UCG is interesting, as well as the possibility of CO_2 storage in the incurred cavity after coal gasification [10,11].

The UCG center at the China University of Mining and Technology in Beijing is also testing UCG in abandoned coal mines. A technical center for UCG has been set up in the University of Beijing, and technical exchange of information on UCG is taking place in the UK [12].

From 1997 through 2003 the Chinchilla project was made in Chinchilla, Queensland, Australia. It was the largest UCG project to date in the western part of Australia. Project layout, design, and operation of the UCG plant were done with the help of Ergo Exergy Technologies Inc., Montreal, Canada (Ergo Exergy) [13].

UCG technology points to the need to know the geology, hydrogeology, and tectonic faults of the deposit and its surroundings. These can affect the rock environment around the underground generator.

The current idea of underground coal gasification is pointed out here, also due to the fact that, according to the currently valid calculations, there are a number of reserves that are impossible to be extracted by current technologies or they were recovered with high losses. The new way of using lignite reserves should contribute to reducing the negative impacts on the elements of the environment and increasing the usage of own resources of mineral wealth.

Mining has a real impact on the environment and on the morphology of the whole region. Manifestations of mining activity are of different characters. Depending on the thickness of the overlying rocks and their composition, there is a decline in the terrain above the excavated area. These morphological deformations have various characteristics and manifestations, from simple cracks that close after a few months to extensive ground disturbances such as soil subsidence, landslides, and rifts.

Larger landslides result in the flooding of the area with water. Local wetlands are created, where new habitats with variable representations of plants and animals are present. Local landslide areas are also formed, especially in hillside areas [14].

Undercutting of unstable slopes has been shown to reactivate slope landslides. The slopes of the Vtacnik mountain range are primarily disturbed by landslides. They are characterized by sensitivity to even minimal interventions in the stability system.

The danger of UCG and general effects are summarized in Table 1.

Table 1. Main dangers and adverse effects of UCG technology [15].

Main Dangers	Adverse Effects of UCG Technology
- carcinogenic waste (coal tar) - contaminating layers creating water	- dust and air pollution at the time of production
- the danger of underground explosions	- inhabited areas are not suitable for research
- gas emissions may come to the surface	- delayed redevelopment of the area
- subsidence may occur even after several years	- the threat of un-controlled expansion of groundwater contamination

At present, according to the mining laws, extensive remediation works are being carried out to adjust the affected areas so that they are economically usable and minimize the adverse effects on the environment.

2. Material and Methods

The article addresses the issue of underground coal gasification, using knowledge of geology, hydrogeology, and tectonic faults of the coal deposit. Gasification is a chemical process of converting solid or liquid fuels into gaseous fuels, which takes place in gasifiers (generators, reactors).

The principle of UCG technology is based on the existence of at least two wells (often a series of wells), namely, injection and production wells, drilled into the coal seam. After ignition of the seam, the oxidizer is blown into the injection borehole and low to medium calorific gas is gathered by the production well. In the bearings, chemical reactions similar to those in conventional gasifying generators run. The extracted gas has a diverse quality, which is dependent on the quality of coal, the type of oxidant, and coping with the process [16,17].

The UCG is a transformation of heavy, liquid fuels into gaseous fuels, which occurs in a coal bed. The gasification of coal under high temperatures causes decomposition of organic substance, and tar, gas, cinder, and ashes are produced. During the UCG, controlled burning occurs under the ground in the coal seam [18,19].

As already mentioned, at least two wells are injected into the coal layer (injection and production). Through the injection well, the layer is burned with gasification medium. By the second-production well, the produced gas gets to the surface (see Figure 1) [20,21].

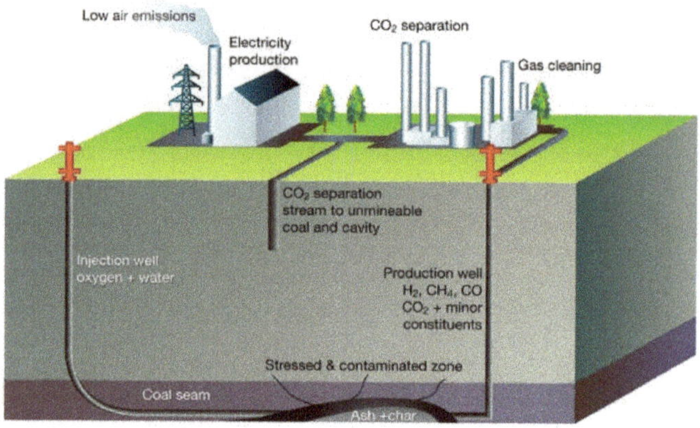

Figure 1. Scheme of UCG technology [22].

The operational parameters of the gasifying agent (air or oxygen) have a significant impact on the UCG process efficiency and could also affect its economic performance [23] and gasification indexes including the syngas quality and yield, by regulating its injection method, mixing ratio, and volume to meet different needs [24,25].

During coal gasification, the action of the gasification medium at high temperatures leads to the decomposition of organic substance. The result is gaseous products, tars, and a solid residue, which is cinders or ash. After cleaning, gaseous products are used for the production of electricity or as a raw material for the production of chemical products.

The extracted gas normally contains more than 80% of methane. The gas composition, particularly the methane content, is decisive for its use. It is commonly used to produce electricity. However, with a high content of methane, over 90%, its use is the same as for natural gas and the gas is pumped into pipelines [26,27].

The underground coal gasification is based on the same principle as classic gasification with the only difference being the place of gasification in the coal seam [28].

UCG technology was verified in Slovakia within the projects, only in laboratory conditions, for several experiments. The experiments differed in the methods of coal storage, the use of different oxidants, and the methods of disposal. Experimental coal gasification in laboratory conditions, which took place during the solution of applied research projects, allowed us to gain knowledge about this process. We analyzed useful but even harmful products of this so-called "clean technology".

Currently, the biggest environmental polluters in the district of Prievidza are Power Plants Slovenské elektrárne, Novaky Power Plant, Zemianske Kostolany Plant, The Novaky Chemical Plant, and Novaky and Upper Nitra Coal Mines [4]. Based on the projects solved so far at the Technical University in Košice dealing with the use of UCG technology in Slovakia, the Mines Cigel and Upper Nitra Coal Mines Prievidza were determined to be the most technically suitable.

From the methodology point of view, we stated for our research specific parameters:

- Geology: geological-structural characteristics of the deposit;
- Hydrogeology: hydrogeological characteristics of the deposit and hydrochemical characteristics of the deposit; and
- Tectonic structure of the deposit.

We stated benefits of this technology based on the accomplished operational trials abroad.

We analyzed the possibility of influencing the rock environment by UCG activities according to specific geological, hydrogeological, and tectonic conditions. Three models of possible situations of behavior of the underground UCG generator on the surrounding environment according to specific conditions were determined.

We proposed measures to eliminate the occurrence of pollution and adverse effects on the environment.

3. Results

3.1. Geological and Hydrogeological Conditions in the Mining Area of the Cigel Mine

There are significant coal deposits in the Upper Nitra Basin. Deposits of caustobiolith (brown coal and lignite) represent a significant raw material potential. Brown coal is situated in the locality of the Handlová deposit and lignite is situated in the Novaky deposit. Three mining areas are registered on two deposits: Cigel Mine, Handlova Mine, and Novaky Mine (see Figure 2).

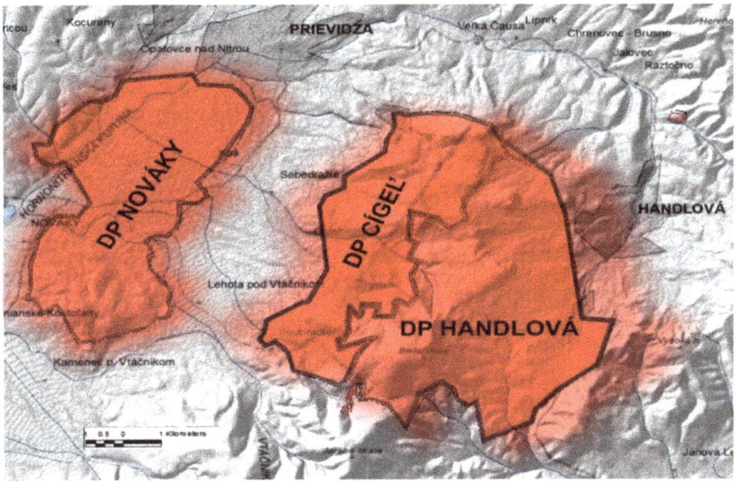

Figure 2. Definition of the territory, studied area: wider surroundings of mining areas (DP) of brown coal deposits Novaky, Cigel, and Handlova [29].

The basic development of the coal basin was formed in the lower and middle Miocene. In this environment, a massive formation of epiclastic rocks originating from the destructive Baden strato-volcano from the southern part of the basin was deposited. The emerging, fluvial-limnic environment created favorable conditions for the formation of rich plant vegetation, from which coal seams were formed. Diagenesis of vegetation layers formed coal seams in the central part of the basin [29].

3.1.1. Geological-Structural Characteristics of the Mining Area Cigel

The Cigel mining area covers the western part of the Handlova Coal Deposit, where the Mine Cigel operates. In the northern part of the mining area is located the VII mining section. The rock mass was built by a complex of Neogen rocks. The top cover was formed by clays, clay stones, and andesite breccia of the highest part of the Sarmat. Part of the filling of the area consists of rocks of volcanic-detrital formation with an average thickness of about 20–50 m. The direct overburden of the productive formation was created by gray Kos-overburden clays. The total thickness of overburden (Kos) clays is variable, from 150 to 230 m.

The productive coal formation consists of several coal and clay layers. It has a total thickness of about 20–55 m. The uppermost coal position is defined as the upper coal seam (h_1). The geological thickness of the layer is about 5–6 m. The total thickness of coal is about 4–4.5 m. The calorific value of coal is about 13.0 $MJ \cdot kg^{-1}$. The compressive strength of the coal deposit ranges from 17 to 21 MPa. The subsoil of the upper layer was formed by the so-called interlinear sandy clays consisting of fine psammitic, illitic clays of varying thickness and strength. Clays have the ability to swell in the presence of water. In the lower part of the formation there is a lower layer-seam (h_2). The calorific value of coal is about 12 $MJ \cdot kg^{-1}$, the compressive strength is from 12 to 18 MPa, and the geological thickness is from 4 to 5 m. The bedrock below the lower layer consists of various tufitic clays, clay stones, and sandstones (stone formations).

Storage conditions of the coal formation in area VII sections are variable, with the lowest points in the central part of the section. In general, the whole area is slightly sloping to the SW. As a result of active subsidence movements of the entire deposit area, sedimentation anomalies occurred in local deposits, resulting in smaller deposits of caustobioliths. In area VII, the section of Cigel Mine has a registered coal thickness of approx. 30–35 m above the upper layer, overlying the coal seam, h_0.

The morphology of the surface is of hilly character with a slope from east to west. The coal formation in the given locality is situated at a depth of 190 to 430 m. After the exploitation work, various modifications of tensile cracks appeared on the surface (individual cracks are up to 1.0 m wide and about 20 m long), which closed after a certain period and left a mediumly devastated surface (smaller sliding areas, disturbed original vegetation, surface defects, and others) [30].

3.1.2. Tectonic Structure of the Deposit

Endogenous processes created a complex tectonic structure consisting of subsidence and reverse fault defects.

The Handlova coal deposit represents a complex, asymmetrically developed ridge with different tendencies of fractures, which are approximately in the NE-SW direction. The prevailing opinion is that the foundation of the tectonic structure of the deposit has a very close connection with the volcanic activity of the Vtacnik mountain range.

The basic network of tectonic fault forms elongated lenticular blocks of various lengths and widths from a coal deposit. The vertical displacement between the individual blocks is very variable and reaches values from decimeters to several tens of meters. The predominant type of tectonic structure is a declining tectonic fault.

The genesis of subsidence failures is evident in the fact that, in areas with volcanic processes, different tensions arise in the foundation, but also in the higher parts of the rock massif. As a result of these unequal tension states, some parts of the mountain massif are displaced (raised, fall) in some areas, which can evoke various disjunctive surfaces in the fragile coal and plastic clay layers.

By penetrating volcanic bodies into the base rocks, they can cause various vertical movements of the rock layers, thus creating a basis for tectonic rock failure. These areas of instability can very often be combined with the gravitational descent of the broken blocks and can form a combined system of breakage in the rock layers.

The inclination of the disjunctive surfaces varies according to the character of the rock layer from 15° to 60°. Some tectonic faults may be of pre-sediment origin, as they can only be identified in the base rocks without penetrating the productive layers. The second groups of tectonic structures are tectonic faults in the coal seam itself. Another group of disjunctive lines are those that disrupt the entire productive complex of strata (from the subsoil through the coal seam to the overlying clays).

Important knowledge about the genesis of tectonic defects was obtained from the area of coal seam bifurcation. Another type of tectonic fault was found in the given locality, where one tectonic line tends to disrupt the coal seam twice, but with opposite inclinations of disjunctive surfaces (so-called paired tectonic faults).

A genetically separate part of the tectonic fault is the reverse fault tectonics, which occur in areas VII and VI of Section B. It is not possible to identify the origin and development of shears and dips. The genesis of the reverse fault in the northern region of the deposit is explained by the active penetration of magma into the productive complex of strata. The tangential component of the pressure vector induced by volcanic processes caused shifts in the upper part of the coal complex of the strata. In the southern part of the deposit there is a shear structure that breaks the entire coal formation. This rearrangement system has genesis in the area of the plutonic body, which is located at the southern boundary of the coal deposit. The guide length depends on the intensity of the stress forces [30].

3.1.3. Hydrogeological Characteristics of the Deposit

The mine "Bana Cigel" mining area belongs to the Nitra river basin. The most important streams flowing through the deposit are Mostenica, Ciglianka, Takov, and Mraznica, which flow into Handlovka. The productive complex of strata is located above the erosive base of the Nitra River with an elevation of about 250 m above sea level.

The area of the entire mining area and its surroundings is actually an infiltration area. Atmospheric precipitation, which seeps into the rock environment and penetrates the volcanic-detrital formation to an impermeable layer of overlying clays, which in an intact state form an impermeable barrier. In the area of the absence of overlying clays, the impermeable layer forms a coal seam. The general inclination of the impermeable subsoil of the volcanic-detritic formation directs the underground filtration flow, which does not run linearly but adapts to the relief of the overlying clays formed by erosion. In addition, it is also affected by the position of individual tectonic blocks.

Streams forming local erosive bases located above the deposit are an important factor in the saturate of the overburden.

Overall, we can determine the following causes of overburden flooding:

(1) permanent underground flow toward the erosive base of the Nitra River,
(2) streams that form local erosive bases and water supply volcanic-detritic formation,
(3) water supplying the volcanic-detrital formation through a series of tectonic faults oriented approximately perpendicularly to the direction of the filtration flow.

In their natural state, overlying rocks are drained by various types of springs and surface flows. However, the established natural hydrogeological regime is significantly affected by mining activities. Drainage and pumping of water from the underground creates new artificial drainage points, which cause changes in the flow of groundwater.

The geological structure of the overburden and the height position of the seam zone in the mining areas of the mine "Bana Cigel" and Handlova conditioned, from the point of view of mining safety, the need to drain mainly rocks of the overlying horizon of volcanic-detritic formation and andesite.

The underlying water bearing in VII section has only local significance. In this area there are layers of tuffites, the irrigation of which is tied to the near infiltration area (Handlova, Morovno Prievidza). The tuffites continue toward the so-called negative zone to the subsoil of the Novaky deposit, where they are intensively drained.

In relation to the seam zone, three hydrogeological units are separated on the deposit: overlying, inter-layer, and underlying. The overlying irrigated horizon with a maximum thickness of up to 400 m is represented by Quaternary sediments, products of overlying volcanism, volcanic detritic formation, and the overlying formation of overburden. The Quaternary is not essential from the point of view of deposit hydrogeology. The products of overlying volcanism are tuffites, tuffo-breccia, and especially andesites of various types. They predominate in the southern part of the deposit, where they reach up to 350 m and here their thickness increases in the east–west direction.

The groundwater of neovolcanics is bound to more intensively cracked parts of volcanic massifs, to broken fault lines, and to their contact with low-permeable positions of volcanoclastics. The filtration coefficient (k) of andesites has a wide range: 1×10^{-7} m·s^{-1} to 1.58×10^{-3} m·s^{-1}. Cracked rock with signs of hydrothermal transformation, with the occurrence of tectonic zones at the base, predetermines the intensive replenishment of the seam zone with water. Individual hydrogeological boreholes, whether surface or mining, are not, in most cases, goundwater levels related, and flow occurs in more locally permeable areas and positions. In most cases, groundwater levels are not continuous between individual hydrogeological boreholes, whether surface or mining, and the flow occurs in more locally permeable positions.

The volcanic-detrital formation is petro graphically and granulometrically very diverse. It is a rock complex characterized by great heterogeneity. It is built of gravel conglomerate sands, clays with sandstones, and tuffits, and tuffites are sporadically represented. The clays are sandy in places. In contact with water they quickly change their consistency from a solid state to a plastic one.

In gravel, conglomerate was found, k from 3.8×10^{-5} to 5.7×10^{-8} m·s^{-1}, and it had more cracked than porous permeability. Pyroclastic-tuffitic breccia tuffaceous-conglomerate has a verified k of 1.97×10^{-7} m·s^{-1} and a combined fissure-porous permeability.

In general, we can speak of a complex water-bearing strata system. The storage part of the system consists of several water-bearing collectors, in which the level is either free or tense. The thickness of the clay to the clay layer separating the seam zone from the volcanic-detritic formation is important for the safe operation of the deposit.

At the southern, eastern, and northwestern edges of the deposit, erosion occurred and there, in direct overburden of seam level, is placed a volcanic-detritic formation or, more precisely, andesite. Overlying clays and clay stones serve as passive hydrogeological protection and, from this point of view, it is important that they are intact and non-disintegrating in contact with water or, more precisely, they stay waterproof.

The inter-seam, water-bearing assise is developed locally. At the bifurcated seam its thickness is max. 32 m. Inter-seam sediments are represented by silica sands, sandy clays, clays, and tuffites with k_f from 6.5×10^{-8} to 9.82×10^{-6} m·s^{-1} [30].

3.1.4. Hydrochemical Characteristics of the Deposit

The chemical composition of groundwater is influenced by the time of water circulation in the rock, geochemical reactions during the mutual mixing of different types of water, and changes in rock environments, temperature, and pressure conditions during water flow. The chemical composition of water is, therefore, variable over time, not only by the action of primary genetic factors but also secondarily by human activities.

Based on physicochemical analyzes, we can define two basic types of water:

(1) the basic calcium bicarbonate type with a transition to the calcium sulphate and magnesium sulphate types, which is characteristic of Quaternary rocks, overlying-overburden volcanism, volcanic-detrital formation, and aged water;
(2) the basic sodium bicarbonate type, which is characteristic of the waters of inter-seam sands and the underlying assise but with higher mineralization [30].

3.2. *Advantages of UCG Technology*

Most countries with large coal reserves have almost 85% of known coal reserves, which cannot be extracted using known mining methods. UCG technology can be the solution to take full advantage of this valuable resource. Many experts argue that this could double the availability of coal worldwide, which is currently expanding the number of UCG projects.

The UCG principle is based on the existence of a minimum of two wells (more often a series of wells), specifically injection and production wells, drilled into the coal seam. After igniting the seam, an oxidizer is injected into the bearing through an injection well and low to medium calorific gas is obtained by production. Chemical reactions in the deposit take place similarly to conventional gasification generators. The obtained gas has different quality and depends on the quality of coal, type of oxidizer, process control, etc. [28].

UCG technology has a less negative impact on the environment because all the coal remains underground and there are fewer emissions and fewer surface footprints because no surface gasification is needed and the gas is processed to remove harmful particles, including CO_2 capture. This process is safe and economical, which also meets the requirement of secure gas supply for domestic and industrial use. Hence, the potential environmental concerns related to UCG need to be addressed and understood to allow for its commercialization [31].

According to operational experiments abroad, the benefits of this technology are summarized in the following points:

- low operating costs of the entire technological process,
- minimal surface changes above or around the deposit during and after mining,
- minimal danger to operating personnel,
- no ash from surface combustion, as in surface gasification,
- significantly lower CO_2 emissions, and
- no surface or underground contamination related to technological processes that have yet been recorded and only minimal reclamation is required after mining [28].

4. Discussion

4.1. Possibility of Influencing the Rock Environment by the Activities of UCG "Underground Coal Gasification"

The possible influence of the rock environment in the vicinity of the underground generator results from the following assumptions:

(1) Contaminants remain in a movable form in and around the generator.

The process of gasification, heating while there is absence of air, proceeds as follows. At temperatures around 100 °C, part of the water is released as well as absorbed gases. Up to 300 °C, a substantial part of the water and a certain amount of gases, mainly carbon dioxide, are released from the brown coal, which releases, in particular, oxygen, a very small amount of nitrogen, and carbon monoxide. By further heating in the range of 300–350 °C, the splitting of bound water and oxygen continues and flammable gases (methane) begin to appear. Only ballast, non-flammable substances are released to this temperature.

Above 350 °C, the decomposition of the carbonaceous component occurs and flammable gases, hydrocarbon vapors, and tars, which occur here in the gas phase, begin to be released. The highest production of vapors of hydrocarbons and tars, which, after cooling, give a liquid fraction, can be obtained at a temperature of about 600 °C. At high temperatures (1000 °C) most of the product is released in the form of gas and their amount decreases rapidly after exceeding a temperature of 800 °C. In terms of solids, a porous residue is formed in connection with the release of gas and vapor, which, if the heating temperatures did not exceed about 600 °C, is referred to as low-temperature coke (semi-coke), when heated to temperatures higher (up to 1000 °C and more) than coke. Higher temperatures cause coal to decompose, resulting in a gaseous fraction and a solid residue [25].

By cooling these gases and vapors, a condensate is formed, which consists of a hydrocarbon and an aqueous fraction. Their chemical character is identical to the fractions of combustibles, which we mentioned as part of the composition of coal. Thus, in hydrocarbon fractions it is the same as in oil: gasoline, diesel, paraffins, light and heavy oils, and asphalt materials. Furthermore, there are aqueous fractions, soluble compounds that are formed by the thermal decomposition of coal. It is mainly ammonia, a certain amount of sulfur substances, and a wide range of organic compounds such as phenols, ketones, and other polar substances. Pure ammonia and nitrogen fertilizers can be obtained by treating ammonia waters. [32] In the period from 2007 to 2010, experiments with coal gasification were carried out at the Technical University in Kosice as part of a grant project. Figure 3 shows the temperatures that occurred during the experimental coal gasification.

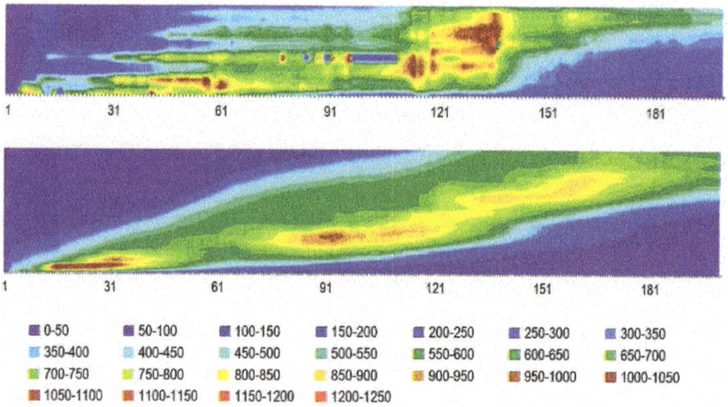

Figure 3. Development of temperatures after the cross section of the generator in time and in laboratory conditions, 1–181 time in hours. The perpendicular axis shows the length of the generator, 0.3 m to 2.6 m [33].

The Environmental Protection Agency (U.S. EPA) has determined the 16 PAHs that were located in the tar. Hydrocarbons are the following: acenaphthene, phenanthrene, anthracene, fluoranthene, pyrene, benzo (a) anthracene, chrysene, benzo (b, k) fluoranthene, benzo (a) pyrene, dibenzo (a, h) anthracene, benzo (g, h, i) perylene, and indeno (1,2,3-c, d) pyrene. Samples contained other pollutants and volatile compounds: TOC and BTEX [34,35].

During the experiments of brown coal gasification in the mentioned two generators in laboratory, tar samples were taken and analyzed in an accredited laboratory [36]. Table 2 shows a chemical analysis of the individual components of tar from about 650 kg of brown coal, and they were compared with the threshold values for synthetically produced pollutants. According to [33], in mine "Bana Cigel", the amount of reserves for underground gasification is about 200,000 t.

Table 2. Values of pollutants in tar samples (NPEC IR, TOC, BTEX, PAH) [28,37] and authors.

Experiment	1	2	Limit Values for Syntheticaly Produced Polluting Waters [μg/L]
Indicator	**Value [μg/L]**	**Value [μg/L]**	
NPEC IR	1,324,000	144,900	
TOC	2,824,000	22,656	
Benzén (BTEX)	3.80	393.90	0.75
o-xylén (BTEX)	3.00	41.00	
m,p-xylén (BTEX)	4.75	76.20	312.5
Toluen (BTEX)	3.20	199.90	437.5
Acenaftén (PAH)	3.08	1022.19	
Acenaftylén (PAH)	2.36	3766.27	
Antracén (PAH)	2.61	880.49	
Benzo(b)flourantén (PAH)	0.00	29.67	
Benzo(a)antracén (PAH)	0.34	358.28	
Benzo(k)flourantén (PAH)	0.00	15.64	
Benzo(g.h.i)perylén (PAH)	0.00	11.58	
Benzo(a)pyrén (PAH)	0.00	44.35	0.00625
Dibenzo(a,h)antracén (PAH)	0.00	5.48	
Fenatrén (PAH)	16.72	2299.92	
Flourantén (PAH)	3.20	1036.90	
Flourén (PAH)	4.92	1162.52	
Chryzén (PAH)	0.38	363.43	
Indeno(1,2,3-c,d)pyrén (PAH)	0.00	20.83	
Naftalén (PAH)	1.77	2894.26	
Pyrén (PAH)	2.26	660.04	
∑ PAH	41.102	14,784	0.0625
Amount of tar [liter]	21.8	10	

In conversion, with a tar production of 31.8 L/650 kg, it represents a total tar production of 9800 L. Such an amount of tar poses a threat of contamination to the rock environment. The use of tar as a gasification product is expected in the chemical industry and, therefore, it is necessary that most of its components are removed from the generator, preferably in gaseous form. By controlling the gasification process, it is necessary to ensure that the condensation of the gaseous components of the tar does not take place in the generator. If this requirement is met, the risk of contamination of the rock environment will also be reduced [38].

High levels of tar cause carcinogens that are dangerous to living organisms.

(2) Contaminants in the generator and in its surroundings get into motion and water enters the generator.

Based on the characteristics of the rock environment given in the previous part 3 it can be stated that in the overburden of productive layers (coal seams) is placed the detrital-volcanic formation, which is water bearing. The productive layer is separated from the detrital formation by a layer of clays with good isolation properties (the filtration coefficient has low values).

Overburden clays and clay stones serve as passive hydrogeological protection and, from this point of view, it is important that they are intact and non-disintegrating in contact with water. The thickness of the clay layer separating the seam zone from the volcanic-detrital formation is important for the safe implementation of the gasification process. Equally important is the tectonic dysfunction that extends into the clay formation because there is a risk of water penetrating the generator and moving the contaminants.

(3) There are communication paths for moving contaminants out of the generator.

As in the previous point, based on the characteristics of the rock environment, it can be stated that the isolating layer of clays forms an impermeable barrier to the exit of contaminants from the generator. However, due to the existence of tectonic lines, the escape of contaminants cannot be completely ruled out.

After analyzing the assumptions of the impact on the rock environment, it is possible to draw the following partial conclusions to minimize the adverse effects on the environment:

(1) Before carrying out the gasification process, the site in question must be assessed geologically and hydrogeologically. The risks of rock and environmental contamination have to be assessed.
(2) To minimize contamination of the rock environment, it is necessary to focus on reducing the amount of contaminants left in the generator.
(3) After the end of the activity in the generator, reduce the content of residual pollutants.
(4) Carry out the injection of the generator premises in such a way that the source of pollutants is modified (physico-chemical), in such a way to limit the mobility of contaminants and their ability to pass into the transport medium.

4.2. *Proposal of Measures to Eliminate Pollution*

4.2.1. Rinsing with Water

In the third and final phase of gasification (Figure 4), the boreholes are liquidated. In case of possible contamination, it is recommended to rinse the cavity with water before disposing of the underground generator and boreholes, in order to minimize possible contamination of the environment and water around the resulting cavity. Potentially contaminated water can be treated in a surface sewage disposal plant. At the end of the process, long-term monitoring of the created cavity and its surroundings is necessary [28].

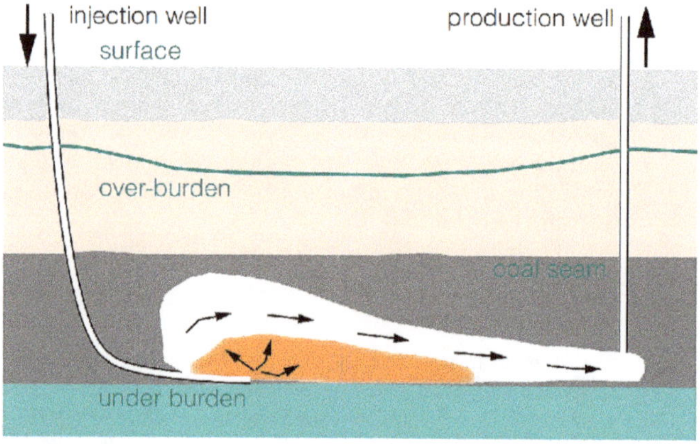

Figure 4. The boreholes of gassification [28].

4.2.2. Grouting with Subsequent Solidification and Stabilization of Materials in the Generator

One of the possibilities of filling boreholes and cavities after gasification is the principle of secondary use of waste for solidification and stabilization. The basic precondition for implementation is to design such an injection mixture that meets the requirements for:

- suspension stability, good penetration into the environment, and good pumping.
- leachability of harmful substances

The penetration ability can be assessed using Darcy's law and the filtration coefficient k.

$$k = \frac{Q.L}{S.\Delta h}\left[\mathrm{m.s^{-1}}\right] \tag{1}$$

where:

Q is flow of penetrating suspension [$m^3.s^{-1}$],
L is distance between places A and B,
S is the area over which the flowing suspension flows [m^2], and
Δh is the pressure difference at the point of inflow and outflow of the suspension [m] [39].

- leachability of harmful substances.

The basic criterion in assessing the impact of mining and construction materials on the environment is the assessment of the leachability of harmful substances from the filling mixture into mining waters. In the natural environment, leachability is a function of the physical and chemical properties of the base mixture and the hydrogeological and geochemical characteristics of the place where it was used for the mining-building material.

The basic factors influencing leachability include:

- surface area of the base mixture (granulometric compound of waste) and its permeability,
- the chemical composition of the solution in which the leaching takes place,
- pH of the leaching solution,
- leaching time,
- leaching temperature, and
- liquid to solid phase ratio.

Due to the usability of the solidificate, cementation methods, which have been used abroad for many years for the disposal and utilization of waste containing mainly heavy metals, are the most suitable for mining conditions. These methods are based on the fixation and immobilization of pollutants in the silicate matrix. It is a physico-chemical treatment of waste by homogenization with suitable components so that no pollutants are released into the natural environment [40].

As with every product, also base mixtures have to meet the criteria for chemical and physical properties that determine the utility value of the product. These criteria can be divided into basic groups:

- assessment in terms of specific use,
- evaluation from the point of view of transport and injection technology, and
- assessment of the impact of mining material on the environment.

Whether the base mix-mining building materials meet the abovementioned criteria has to be verified by tests under such conditions as they are used at the place after solidification. The leachability test has to be carried out in the state in which it arises after use and after stabilization and solidification (not a ground sample), as in the case of a construction product.

In order to prepare base mixes meeting the above criteria, we need to know the chemical and physical properties of the waste, as these determine the final formulation of the mix, which must guarantee that chemical substances are stabilized in the production and subsequent solidification process [39]. It is the method that, by its nature, meets the requirements of environmental protection and labor safety [41,42].

5. Conclusions

The district of Prievidza is one of the most industrially developed and urbanized regions of Slovakia, with a predominance of fuel, energy, mining, and chemical industries. This concentration of industry was reflected in the state of the environment, the quality of which ranked the district among the districts with the most polluted environment in Slovakia. As a result, part of the district was included in one of the eight congested areas of Slovakia–Ponitrianska congested area [33].

In the case of using the so-called clean technology, underground coal gasification, the article draws attention to the possible impact of pollution, taking into account the geological, hydrogeological, and tectonic conditions in the selected locality of mine "Bana Cigel". Attention was focused on pollution from underground coal gasification in situ, taking into account the amounts of gasified coal based on sampling after simulated gasification and chemical analyses of tar. The question remains how these tars will affect the groundwater and surface water in this area, which has long been burdened by the mining and chemical industries.

Based on the analysis of the produced gas (syngas), a mixture of tar and water, the article pointed out the energy properties of the products as well as hazardous substances that may endanger the environment. The laboratory gasification of coal, which took place on the surface, within the solved projects, in the proposed generators, made it possible to monitor individual useful but also harmful products of this so-called clean technology.

The proposed measures to mitigate adverse effects after in situ gasification aim to minimize these adverse effects. It was pointed out here to select suitable materials for the implementation of filling/grouting of spaces that will remain empty after gasification and around the wells.

The presented results and the difference between our experimental data and data acquired in some other studies indicate the need for further research.

Author Contributions: Conceptualization, S.Z. and E.S.; methodology, E.S. and A.S.; validation, A.S. and G.W.; formal analysis, E.S. and A.S.; investigation, E.S. and S.Z.; resources, E.S. and A.S.; data curation, G.W.; writing—original draft preparation, E.S. and S.Z.; writing—review and editing, G.W.; visualization, A.S. and G.W.; supervision, E.S. and A.S.; project administration, E.S.; funding acquisition, E.S. All authors have read and agreed to the published version of the manuscript.

Funding: This research received no external funding.

Institutional Review Board Statement: Not Applicable.

Informed Consent Statement: Not Applicable.

Data Availability Statement: Not Applicable.

Acknowledgments: This contribution is the result of the project of APVV-0582-06 Underground gasification of coal by thermal decomposition and Hornonitrianske bane Prievidza a.s., Slovakia.

Conflicts of Interest: The authors declare no conflict of interest.

Abbreviations

CO_2	carbon dioxide
BTEX	benzene, toluene, ethylbenzene, xylene
UCG	underground coal gasification
TOC	total organic carbon
PAH	polycyclic aromatic hydrocarbons
NPEC IR	non-polar etxtractable compounds spectrophotometric method in the infrared region of the spectrum

References

1. Uznesenie Vlády SR č. 47/2010. 2010. Available online: https://www.economy.gov.sk/uploads/files/4KU93HLr.pdf (accessed on 21 April 2021).
2. Kacur, J.; Durdan, M.; Laciak, M.; Flegner, P. Impact analysis of the oxidant in the process of underground coal gasifica-tion. *Measurement* **2014**, *51*, 147–155. [CrossRef]
3. Kapusta, K.; Wiatowski, M.; Stanczyk, K. An experimental ex-situ study of the suitability of a high moisture ortho-lignite for underground coal gasification (UCG) process. *Fuel* **2016**, *179*. [CrossRef]
4. Hornonitrianske Bane Prievidza, a.s. Available online: https://www.hbp.sk/index.php/sk/spolocnost/tlacovy-servis/spravy2019 (accessed on 21 April 2021).
5. Návrh Energetickej Politiky. 2015. Available online: http://www.rokovania.sk/Rokovanie.aspx/BodRokovaniaDetail?idMaterial=23993 (accessed on 21 April 2021).
6. EEA (European Environment Agency). Renewable Energy in Europe 2017. In *Recent Growthand Knock-On Effects*; EEA Report No. 3/2017; EEA: Copenhagen, Denmark, 2017.
7. Rybar, R.; Kudelas, D. Energy sources—Division and explication in contects. *Acta Montan. Slovaca* **2007**, *12*, 269–273.
8. Kudelas, D.; Tausova, M.; Taus, P.; Gabaniova, L.; Kosco, J. Investigation of operating parameters and degradation of photovoltaic panels in a photovoltaic power plant. *Energies* **2019**, *12*, 3631. [CrossRef]
9. Lukac, L.; Rimar, M.; Variny, M.; Kizek, J.; Lukac, P.; Jablonsky, G.; Janosovsky, J.; Fedak, M. Experimental investigation of primary DE-NOx Methods Application Effects on NOx and CO Emissions from a Small-Scale Furnace. *Processes* **2020**, *8*, 940. [CrossRef]
10. Schiffrin, D.J. The feasibility of in situ geological sequestration of supercritical carbon dioxide coupled to underground coal gasification. *Energy Environ. Sci.* **2015**, *8*, 2330–2340. [CrossRef]
11. Khan, M.; Mmbaga, J.P.; Shirazi, A.S.; Trivedi, J.; Liu, Q.; Gupta, R. Modelling Underground Coal Gasification- A Review. *Energies* **2015**, *8*, 12603–12668. [CrossRef]
12. Ding, T.; Ning, Y.; Zhang, Y. The Contribution of China's Outward Foreign Direct Investment (OFDI) to the Reduction of Global CO_2 Emissions. *Sustainability* **2017**, *9*, 741. [CrossRef]
13. Burton, E.; Friedmann, J.; Upadhye, R. *Best Practices in Underground Coal Gasification*; Lawrence Livermore National Laboratory: Livermore, CA, USA, 2014.
14. Tianhong, D.; Zuotang, W.; Limin, Z.; Dongdong, L. Gas production strategy of underground coal gasification based on multiple gas sources. *Sci. World J.* **2016**. [CrossRef]
15. Ekneligoda, T.C.; Marshall, A.M. A coupled thermal-mechanical numerical model of underground coal gasification (UCG) including spontaneous coal combustion and its effects. *Int. J. Coal Geol.* **2018**, *199*, 31–38. [CrossRef]
16. Brown, K.M. In situ coal gasification: An emerging technology. *J. Am. Soc. Min. Reclam.* **2012**, *1*, 103–122. Available online: https://www.asrs.us/Portals/0/Documents/Journal/Volume-1-Issue-1/Brown-CO.pdf (accessed on 2 May 2021). [CrossRef]
17. Mocek, P.; Pieszczek, M.; Swiadrowski, J.; Kapusta, K.; Wiatowski, M.; Stańczyk, K. Pilot-scale undergroundcoal gasification (UCG) experiment in an operating Mine "Wieczorek" in Poland. *Energy* **2016**, *111*, 313–321. [CrossRef]
18. Uppal, A.A.; Bhatti, A.I.; Aamir, E.; Samar, R.; Khan, S.A. Control oriented modeling and optimization of one dimensional packed bed model of underground coal gasification. *J. Process Control* **2014**, 1. [CrossRef]
19. Doucet, D.; Perkins, G.; Ulbrich, A.; du Toit, E. Production of power using underground coal gasification. *Energy Sources Part A Recovery Util. Environ. Eff.* **2016**, *38*, 3653–3660. [CrossRef]
20. Lin, X.; Zuotang, W.; Wengang, H.; Guojun, K.; Xuefeng, L.; Peng, Z.; Jianhua, W. Temperature field distribution of burnt surrounding rock in UCG stope. *Int. J. Min. Sci. Technol.* **2014**, *4*, 573–580.

21. Bhutto, A.W.; Bazmi, A.A.; Zahedi, G. Underground coal gasification: From fundamentals to applications. *Prog. Energy Combust. Sci.* **2013**, *39*, 189–214. [CrossRef]
22. Scheme of UCG Technology. Available online: https://www.google.com/search?q=UCG+ (accessed on 21 April 2021).
23. Perkins, G.; Vairakannu, P. Considerations for oxidant and gasifying medium selection in underground coalgasification. *Fuel Process Technol.* **2017**, *165*, 145–154. [CrossRef]
24. Huanga, W.-g.; Wang, Z.-t.; Tian-hong Duand, T.-h.; Xin, L. Effect of oxygen and steam on gasification and power generation in industrial tests of underground coal gasification. *Fuel* **2021**, *289*, 119855. [CrossRef]
25. Xu, B.; Yang, M.K.; Xing, B.L.; Su, F.Q.; Chen, L.J.; Wang, F.; Zhang, Y.L.; Yi, G.Y. Removal of pollutants from aqueous solutions by coals and residual cokes obtained from simulated underground coal gasification experiments. *Fuel* **2021**, *292*, 120292. [CrossRef]
26. Burchart-Korol, D.; Krawczyk, P.; Czaplicka, K.; Smolinski, A. Eco-efficiency of underground coal gasification (UCG) for electricity production. *Fuel* **2016**, *173*, 239–246. [CrossRef]
27. Białecka, B. Estimation of coal reserves for UCG in the upper silesian coal Basin. *Nat. Resour. Res.* **2008**, *17*, 21–28. [CrossRef]
28. Sasvari, T.; Blišťanová, M. *Možnosti Získavania Energetického Plynu z Uhoľných Ložísk*, 1st ed.; Faculty of BERG, Technical University of Kosice: Kosice, Slovakia, 2007; p. 167.
29. Fazekas, J. Plynonosnost uholnych slojov v Hornonitrianskej panve—Dobyvaci priestor Cigel. *Acta Montan. Slovaca* **2009**, *14*, 57–65.
30. Fazekas, J.; Laukova, I. Geologicka a hydrogeologicka charakteristika loziska v dobyvacom priestore bane Cigel. In *Spravodaj BV 2-3/95*; Spravodaj: Vysna Boca, Slovakia, 1995; pp. 45–54.
31. An, N.; Zagorscak, R.; Thomas, H.R.; Gao, W. A numerical investigation into the environmental impact of underground coal gasi fi cation technology based on a coupled thermal-hydro-chemical model. *J. Clean. Prod.* **2021**, *290*, 125181. [CrossRef]
32. Zelenak, S.; Skvarekova, E. Underground gasification of coal in mining area Cigel, geological and hydrogeological conditions. *Odpadove Forum OZE* **2012**, *7*, 1–11, Praha, 10.
33. *Technicka Univerzita Fakulta BERG v Košiciach, Ustav Riadenia a Informatizacie Procesov, Ustav Geovied, HBP a.s. Prievidza: Podzemne Splynovanie Uhlia Termickym Rozkladom č*; APVV 0582-06; Faculty of BERG, Technical University of Kosice: Kosice, Slovakia, 2008.
34. Laciak, M.; Kostur, K.; Durdan, M.; Kacur, J.; Flegner, P. The analysis of the underground coal gasification in experimental equipment. *Energy* **2016**, *114*, 332–343. [CrossRef]
35. Perkins, G. Underground coal gasification- PartI. Field demontrations and process performance. *Prog. Energy Combust. Sci.* **2018**, *67*, 158–187. [CrossRef]
36. Janoszek, T.; Masny, W. CFD Simulations of Allothermal Steam Gasification Process for Hydrogen Production. *Energies* **2021**, *14*, 1532. [CrossRef]
37. Imran, M.; Kumar, D.; Kumar, N.; Qayyum, A.; Saeed, A.; Bhatti, M.S. Environmental concerns of underground coal gasification. *Renew. Sustain. Energy Rev.* **2014**, *31*, 600–610. [CrossRef]
38. Kapusta, K.; Stanczyk, K.; Wiatowski, M.; Checko, J. Environmental aspects of a field-scale underground coal gasification trial in a shallow coal seam at the Experimental Mine Barbara in Poland. *Fuel* **2013**, *113*, 196–208. [CrossRef]
39. Verfel, J. *InjektováníHornin a Výstavba Podzemních Sten [Rock Grouting and Diaphragm Wall Construction]*; MÚS Bradlo: Bratislava, Czech Republic, 1992; p. 551. ISBN 80-7127-043-1. (In Czech)
40. Sofranko, M.; Khouri, S.; Vegsoova, O.; Kacmary, P.; Mudarri, T.; Koncek, M.; Tyulenev, M.; Simkova, Z. Possibilities of Uranium Deposit Kuriskova Mining and Its Influence on the Energy Potential of Slovakia from Own Resources. *Energies* **2020**, *13*, 4209. [CrossRef]
41. Zakon c. 282/2010, Z.z. nariadenie Vlady SR, ktorym sa ustanovuju prahove hodnoty a zoznam utvarov podzemnych vod. *Zbierky Zakonov* **2010**, *110*, 2328.
42. Quosay, A.A.; Knez, D.; Ziaja, J. Hydraulic fracturing: New uncertainty based modeling approach for process design using Monte Carlo simulation technique. *PLoS ONE* **2020**. [CrossRef] [PubMed]

energies

MDPI

Article

Determination of the Extent of the Rock Destruction Zones around a Gasification Channel on the Basis of Strength Tests of Sandstone and Claystone Samples Heated at High Temperatures up to 1200 °C and Exposed to Water

Krzysztof Skrzypkowski [1,*], Krzysztof Zagórski [2] and Anna Zagórska [3]

1 Faculty of Civil Engineering and Resource Management, AGH University of Science and Technology, Mickiewicza 30 Av., 30-059 Kraków, Poland
2 Faculty of Mechanical Engineering and Robotics, AGH University of Science and Technology, Mickiewicza 30 Av., 30-059 Kraków, Poland; zagkrzys@agh.edu.pl
3 Research Centre in Kraków, Institute of Geological Sciences, Polish Academy of Science, Senacka 1, 31-002 Kraków, Poland; a.zagorska@ingpan.krakow.pl
* Correspondence: skrzypko@agh.edu.pl; Tel.: +48-126-172-160

Abstract: This article presents the results of laboratory tests regarding the influence of high temperatures on changes in the strength and structural parameters of rocks that are present in the immediate vicinity of a gasification channel. Sandstone and claystone samples were heated at 300 °C, 600 °C, 900 °C and 1200 °C. Additionally, the heated samples were placed in water for 24 h. The results of the laboratory tests were used in the numerical simulation using RS2 software. The main goal of modeling was to determine the extent of the rock destruction zone around the gasification channel for dry and wet rock masses. In the numerical simulations, three widths of the gasification channel and three ranges of high-temperature impact were modeled. On the basis of the obtained results, it was found that the extent of rock destruction, both in the roof and in the floor, is greater by several percent for a wet rock mass. For the first time, this research presents the effect of water on heated rock samples in terms of the underground coal gasification process. The results of laboratory tests and numerical simulations clearly indicate a reduction in strength, deformation and structural parameters for the temperature of 1200 °C.

Keywords: high temperature; strength and structural parameters of rocks after heating; destruction zone around gasified channel

Citation: Skrzypkowski, K.; Zagórski, K.; Zagórska, A. Determination of the Extent of the Rock Destruction Zones around a Gasification Channel on the Basis of Strength Tests of Sandstone and Claystone Samples Heated at High Temperatures up to 1200 °C and Exposed to Water. *Energies* **2021**, *14*, 6464. https://doi.org/10.3390/en14206464

Academic Editors: Ján Kačur, Milan Durdán and Marek Laciak

Received: 6 September 2021
Accepted: 7 October 2021
Published: 9 October 2021

Publisher's Note: MDPI stays neutral with regard to jurisdictional claims in published maps and institutional affiliations.

1. Introduction

Underground coal gasification is a prospective method for obtaining useful minerals, in particular from deposits that are considered sub-balance. First of all, it is an environmentally friendly method due to the lack of waste generation on the surface [1] and a much smaller number of preparatory excavations [2]. Coal seams which are accessed using unprofitable opencast or underground methods are the subject of particular interest [3]. One of the criteria for underground mining is coal thickness. For example, in accordance with the Polish criteria for the balance of mineral deposits, the minimum thickness of hard coal in the seam together with interlayers up to 0.3 m thick should be at least 1 m for balance deposits and 0.6 m for sub-balance deposits [4]. Underground gasification encounters a number of obstacles resulting from changing geological and hydrogeological conditions [5]. As a result of high temperatures, the solubility of pollutants in water increases and the possibility of their migration to aquifers occurs. Hazardous inorganic pollutants include ammonia and cyanides [6]. During the process, numerous impurities in the form of aromatic organic compounds are formed, including: benzene, toluene, ethylbenzene, xylenes, phenols and polycyclic aromatic hydrocarbons. Additionally, significant amounts of heavy metals may be released from coal and the ashes generated during gasification in volatilization processes, which are favored by the high temperature of the process and the presence of numerous chemical factors [7]. Rock minerals have specific thermal properties. Heat conduction through the minerals is partially absorbed as energy. The heat is absorbed differently by the mineral, depending on the direction of the heat flow in relation to the crystallographic axis. The pronounced directivity in heat conduction in many minerals is similar to the directivity in the refraction of light rays and the coefficient of linear thermal expansion. The thermal conductivity of rocks can be defined as the transfer of thermal energy by the disordered movement of particles from higher to lower temperatures [8]. Tian et al. [9] pointed out that high temperatures lead to micro-cracks and damage rock microstructures. Liu et al. [10] distinguished three stages of temperature propagation in the surrounding rock of a combustion cavity and stated that mechanical properties of coal and rock are determined by its extreme temperature. Perkins [11] indicated that the coal spalling process leads to cavity growth. Min et al. [12] found a relationship between pyrolysis temperature and pore fissures, ranging from rough and porous to relatively smooth. Feng et al. [13] and Deming et al. [14] stated that the gasification reaction occurs on the surface of micropores.

A gasifying agent (oxygen, oxygen together with water vapor, heated air or oxygen-enriched air) and gas can sometimes filter through rocks in an undesirable direction. It is also worth paying attention to the fact that as a result of high temperatures, the rocks surrounding a gasified coal seam change their porosity and permeability [15]. Yavuz et al. [16] found that for carbonate rocks, the bulk density decreased with increasing temperature. Chaki et al. [17] noticed that for granite rock samples, there was an increase in porosity in the temperature range from 500 °C to 600 °C, which is related to the increase in the number of fractures. Tian et al. [18] indicated that for sandstone, the thermal expansion of minerals changes with increasing temperature, which in turn contributes to a change in the microcrack network and the spread of structural damage to rocks. Małkowski et al. [19] found a relationship between high temperatures and the thermal conductivity factor, the value of which for sandstone, claystone and siltstone increase fourteen times at 1000 °C compared to 20 °C. In the gasification channel, oxidation, reduction and pyrolysis zones can be separated [20]. A characteristic feature of the underground gasification process is the fact that each of these reactions takes place at a temperature of several hundred degrees, and even exceeding 1000 °C [21]. Such high temperatures change the strength and post-critical parameters of rocks and rock mass and obviously change the stability conditions underground, which are necessary in the process of underground coal gasification [22]. Otto et al. [23] determined that parameters such as tensile strength, elastic modulus and the linear thermal expansion factor have a direct impact on changes in stresses and strains around the georeactor. Model studies confirm that impact of a range of high temperatures significantly exceeds the boundaries of the gasification channel [24]. In both laboratory [25] and industrial conditions [26], temperatures above 1000 °C may remain in the gasification channel for more than 36 h after ignition. Falsztinskij et al. [27], on the basis of model studies, determined that the maximum range of the temperature field was located above the fire channel; at the edge of the transition of the oxidation zone into the reduction zone with a temperature of 950–1200 °C, the following parameters were found: width of the transition zone, 0.74–1.5 m; height of temperature propagation, up to 9 m from the coal seam perpendicular to the rock stratification.

Despite the significant progress that has been made in both industrial and laboratory work on the underground coal gasification process, there is still little data on the behavior of rocks under the influence of high temperatures which were then exposed to water after this process. Therefore, this article presents laboratory tests regarding sandstone and claystone rock samples which were heated at 300 °C, 600 °C, 900 °C and 1200 °C. Additionally, after being heated, a new series of samples were immersed in water for 24 h and subjected to strength tests. The aim of the research was to compare the results with samples tested at

20 °C. The results of laboratory tests were used in numerical modeling, the aim of which was to determine the extent of the destruction zone around the gasification channel.

2. Preparation of Samples for Testing

Rock samples of sandstone and claystone were collected from the floor of the preparatory roadway in one of the mines of Jastrzębska Spółka Węglowa in Poland, which was closed due to the small thickness of hard coal deposits. The roadway was located at a depth of 900 m. The coal seam was located at a depth of 980 m. The average thickness of the seam in the area covered by the research was 0.4 m. Directly in the roof of the seam there was claystone with a thickness of 0.6 m, above which, sandstone with a thickness of 27 m was deposited. At the floor of the seam, there was a 0.5 m-thick claystone, below which, there was a 22 m-thick sandstone layer (Figure 1a). Rock cores with diameters of 48 mm were taken from the floor of the roadway (Figure 1b,c). The claystone was gray and dark gray in color and showed a solid and orderly structure. Sedimentary structures, parallel and diagonal layering were visible. In the case of texture, the grain skeleton was compact, with grain diameter varying from the very fine sand fraction to the aleurite fraction. The binder was porous and in terms of chemical composition, it was clay–silica. The mineral composition was mainly quartz, clay minerals in various proportions and mica. In turn, medium-grained sandstone was gray with a massive and ordered structure. Macroscopically, it was concise and uniform, without voids and cavities. In the sandstone, sedimentary structures and layering were visible, emphasized by a change in color and a variable grain diameter. The grain skeleton was compact. The sandstone binder was porous and siliceous. In terms of mineral composition, quartz dominated; moreover, there were traces of feldspar, mica and carbonaceous matter that formed streaks.

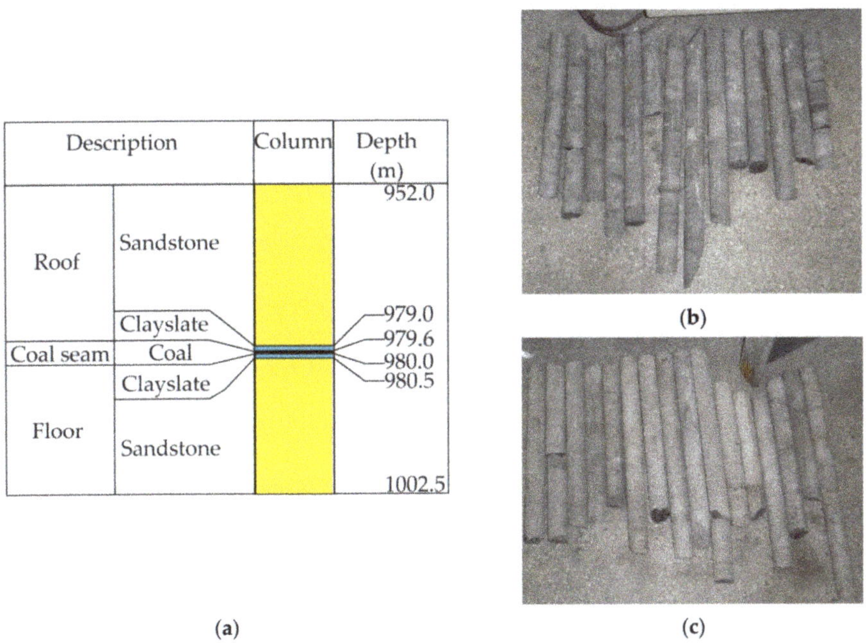

Figure 1. Place of sampling: (**a**) lithological profile of the immediate vicinity of the hard coal seam; (**b**) claystone cores; (**c**) sandstone cores.

The rock cores were cut into cylindrical samples with a height-to-diameter ratio of 2:1. The samples were prepared in such a way that it was possible to apply the load in the direction perpendicular to the stratification (Figure 2a,b). Laboratory tests were divided into several stages. First, the tests were carried out at the temperature of 20 °C. Then, the samples were heated for 24 h at the temperatures of 300 °C, 600 °C, 900 °C and 1200 °C. A laboratory furnace with a maximum heating temperature of 1600 °C was used in the tests (Figure 2c). In the last stage of the tests, a new series of samples were heated at the same temperatures as before, and then the samples were immersed in water for 24 h. The samples were heated in the laboratory of the Faculty of Mechanical Engineering and Robotics, AGH University of Science and Technology in Krakow.

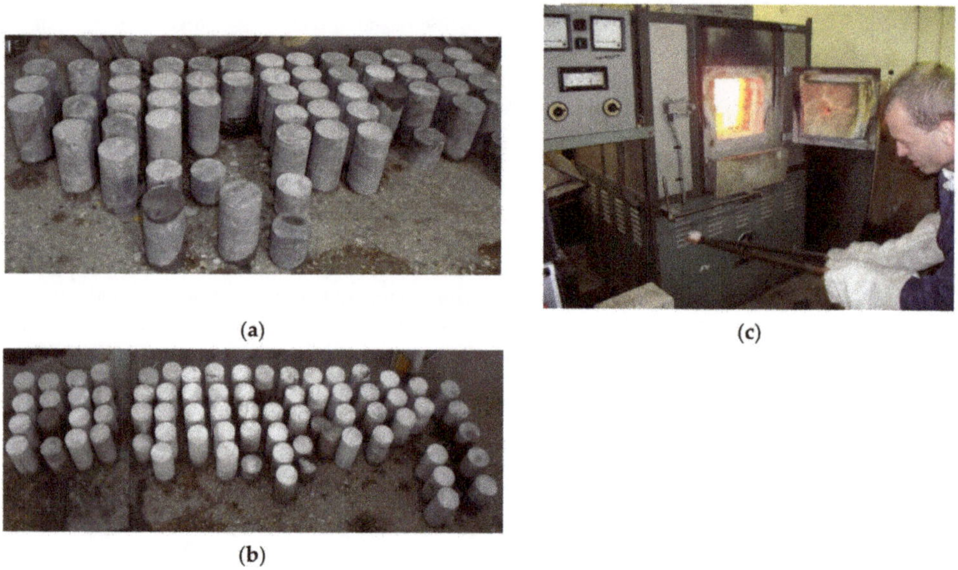

Figure 2. Preparation for tests: (**a**) regular samples of claystone; (**b**) regular samples of sandstone; (**c**) heating in a laboratory furnace.

Research on the influence of water on the strength and deformation parameters of rocks after heating took place due to the fact that the hard coal seam was classified as being a third-degree water hazard zone. On the basis of drilling and exploratory research in the mine, it was found that in the roof and floor of the deposit, there was an aquifer of the fissure and fissure-cavern type, not separated with a sufficient thickness and with a continuous insulating layer from the deposit. In addition, in the part of the rock mass where mining was planned to be carried out, there were water reservoirs containing water under pressure in relation to the floor of these seams.

3. Strength, Deformation and Structural Parameters of Sandstone and Claystone Heated at High Temperatures

Strength tests regarding sandstone and claystone heated at 300 °C, 600 °C, 900 °C and 1200 °C were carried out in a hydraulic press at the laboratory of the Faculty of Civil Engineering and Resource Management at the University of Science and Technology in Krakow. The load measurement was carried out using three strain gauge force sensors, while to determine the vertical deformation, a line encoder with a sampling frequency of 100 Hz was used. In order to determine the horizontal deformations, three electronic sensors were used and spaced 120° apart on the circumference of the sample. The load rate was 0.1 kN/s. After heating, the samples were tested at room temperature, about 20 °C. The load and deformation sensors were connected to the measuring amplifier, which in turn was connected to a computer on which the load–displacement characteristics were monitored on an ongoing basis. Young's modulus was determined at the value of 20–80% of the breaking stress. The density of the samples was determined using the hydrostatic method [28] with the use of a WPS 210/C/1 laboratory balance. Additionally, for the temperatures of 300 °C, 600 °C, 900 °C and 1200 °C, the weight loss was determined. The samples were weighed on a WPT 2 laboratory balance. As a result of the temperatures of 600 °C, 900 °C and 1200 °C, the claystone samples lost their cohesion and divided into irregular pieces. Moreover, several samples were also damaged at the temperature of 300 °C (Figure 3). On the other hand, sandstone samples heated at 300 °C, 600 °C, 900 °C kept their form. Only at the temperature of 1200 °C did numerous cracks appear (Figure 4). The results of the compressive strength, tensile strength, Young's modulus, Poisson ratio, density and weight loss tests are shown in Figures 5–10.

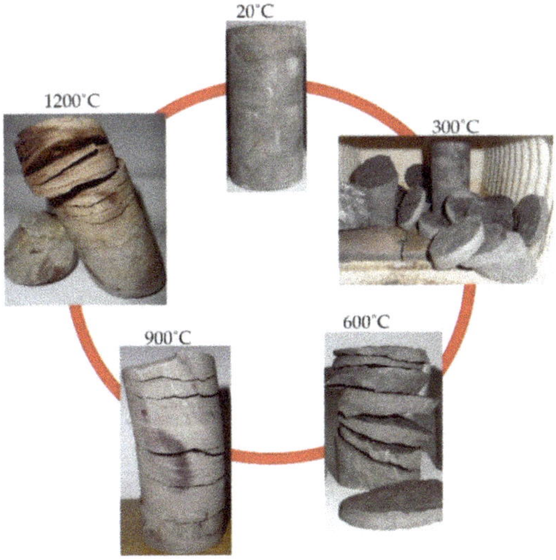

Figure 3. Claystone heated at high temperatures.

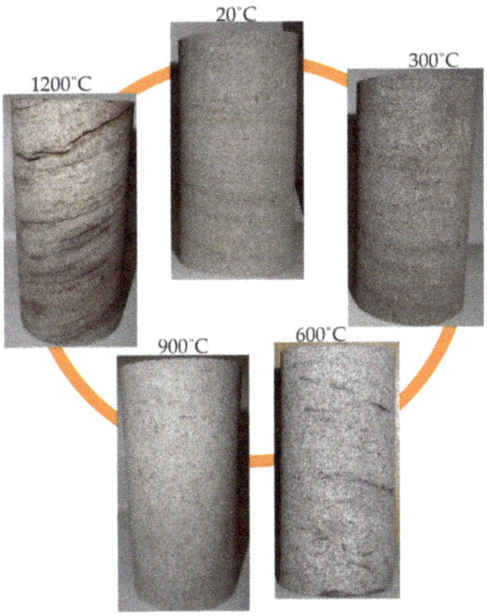

Figure 4. Sandstone heated at high temperatures.

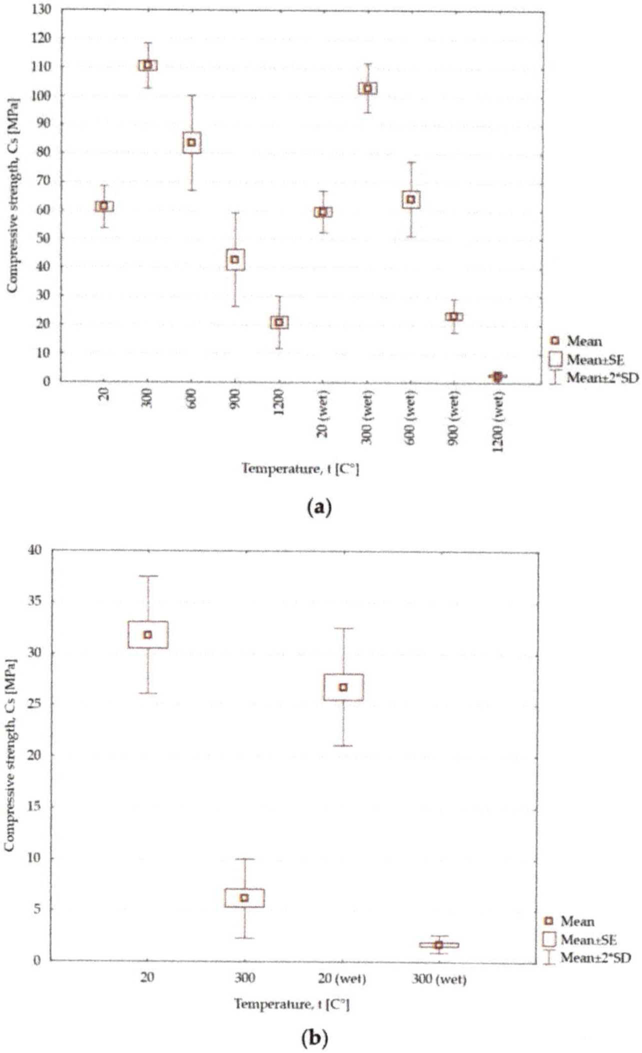

Figure 5. The influence of high temperatures on changes in uniaxial compression strength of dry and wet samples: (**a**) sandstone; (**b**) claystone; SE—standard error; SD—standard deviation.

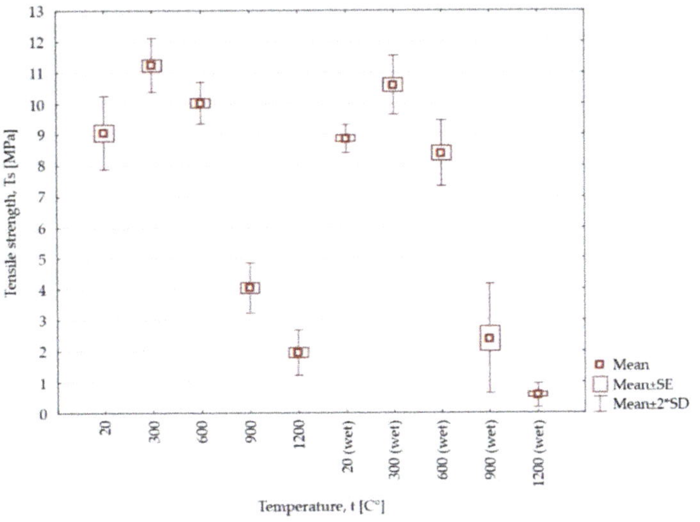

(a)

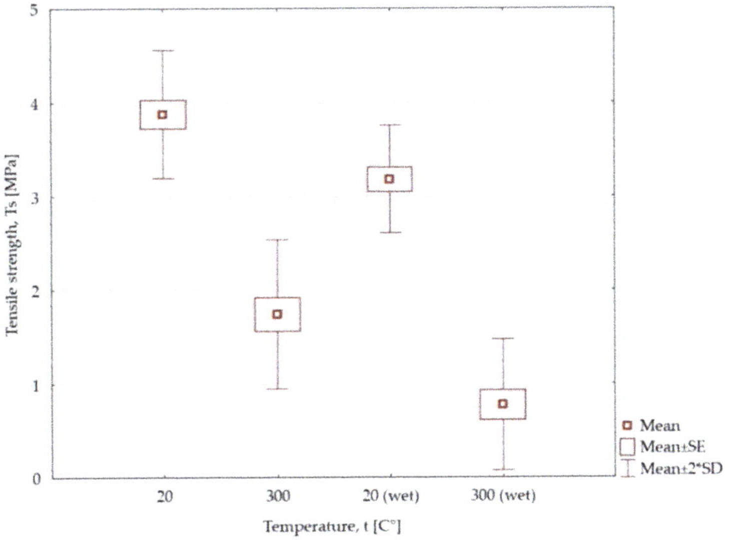

(b)

Figure 6. The influence of high temperatures on changes in tensile strength of dry and wet samples: (**a**) sandstone; (**b**) claystone; SE—standard error; SD—standard deviation.

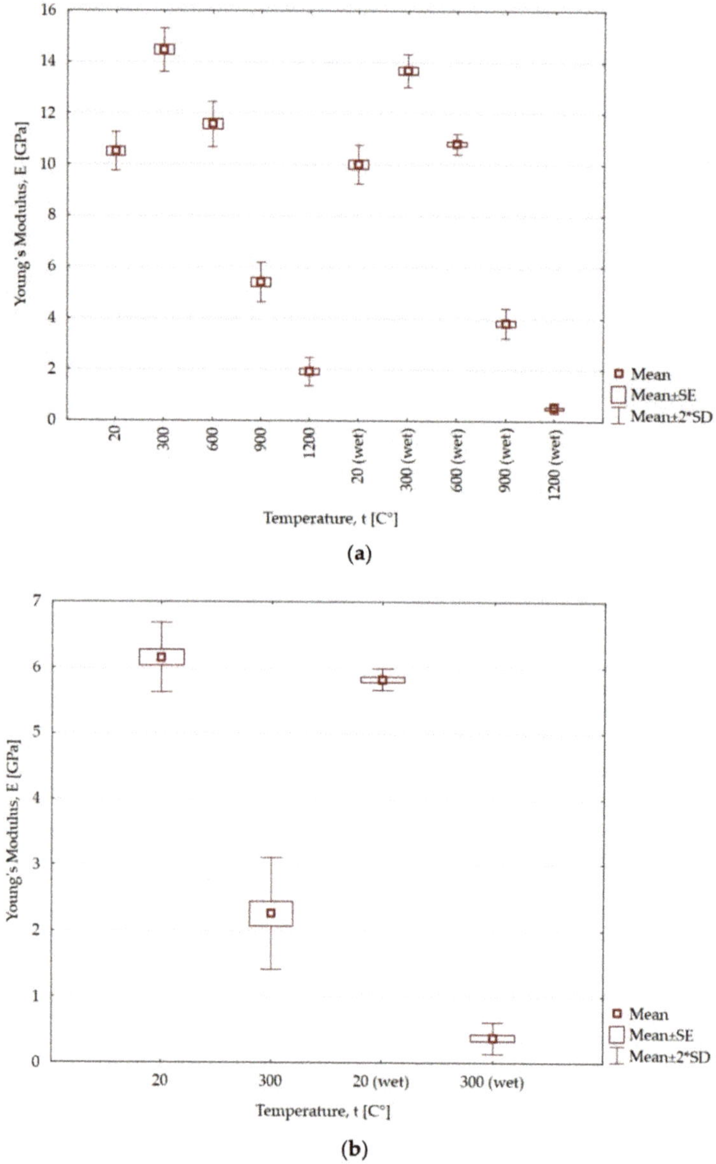

Figure 7. The influence of high temperatures on the changes in Young's modulus: (**a**) sandstone; (**b**) claystone; SE—standard error; SD—standard deviation.

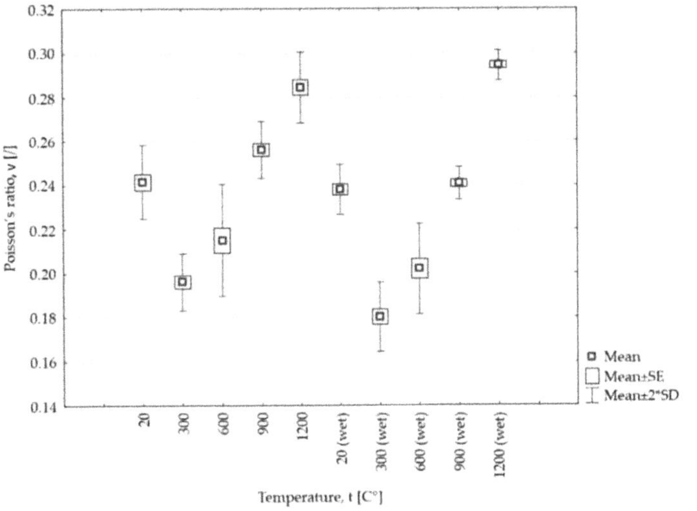

(**a**)

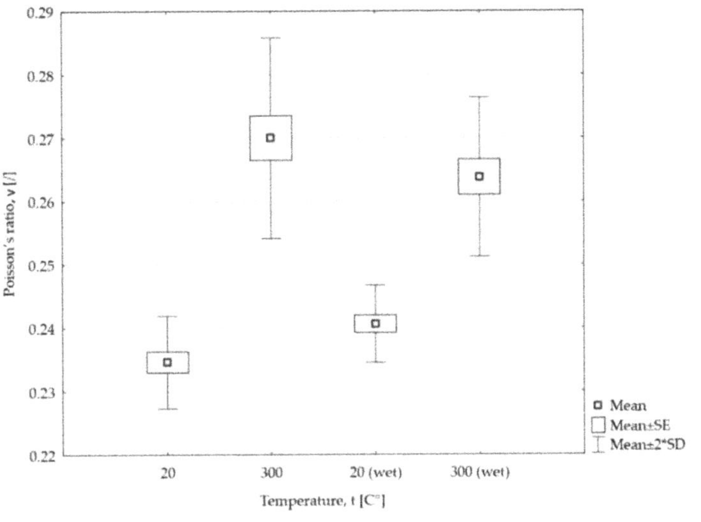

(**b**)

Figure 8. The influence of high temperatures on changes in Poisson's ratio: (**a**) sandstone; (**b**) claystone; SE—standard error; SD—standard deviation.

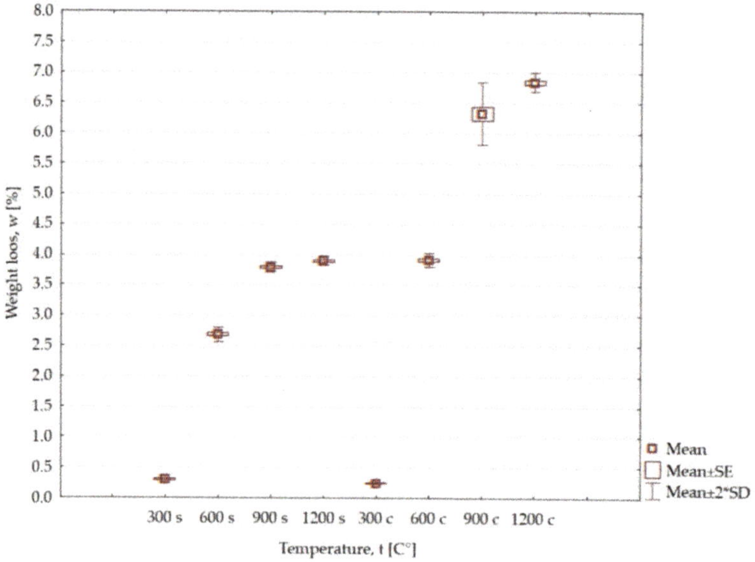

Figure 9. The influence of high temperatures on the weight loss for sandstone and claystone samples: s—sandstone; c—claystone; SE—standard error; SD—standard deviation.

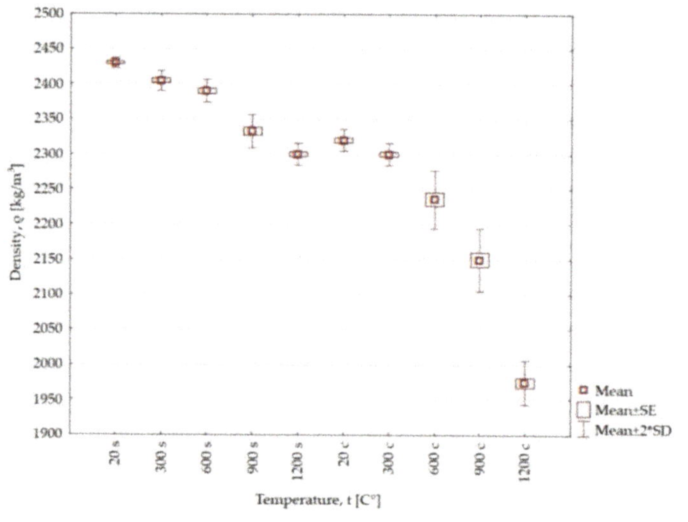

Figure 10. The influence of high temperature on the weight loss for sandstone and claystone samples: s—sandstone; c—claystone; SE—standard error; SD—standard deviation.

The average compressive strength of sandstone samples at 20 °C was 61.2 MPa. Increasing the temperature to 300 °C and 600 °C increased the strength by 80% and 36%, respectively. On the other hand, at the temperatures of 900 °C and 1200 °C, a significant decrease in strength was recorded, by 30% and 65.6%, respectively, in relation to the initial value. For the heated samples, which were then kept in water for 24 h, the range of changes compared to the temperature of 20 °C was an increase of 72% and 7.7% for the temperatures of 300 °C and 600 °C, respectively. The influence of temperature and water caused a drop in strength by 60.7% and 95.7%, respectively, for the temperatures of 900 °C and 1200 °C.

The influence of water on the heated samples caused a decrease in compressive strength by 8%, 28.3%; 30.7% and 30.1%, respectively, for the temperatures of 300 °C, 600 °C, 900 °C and 1200 °C compared to samples that were only heated. In the case of claystone, the influence of the temperature of 300 °C caused a drop in strength by 80.5% and 93.3% for the heated and wet samples, respectively, compared to the initial value. This means that the influence of water contributes to a further decrease in strength by almost 13%.

Tensile strength is one of the basic constants describing the strength properties of rock material, which can be determined using direct and indirect methods in accordance with the recommendations of the International Society for Rock Mechanics and Rock Engineering [29]. Tensile strength tests were carried out using the Brazilian method (transverse compression), in which the tensile force was created in the axial cross-section of a sample, perpendicular to the compressive load. The test specimens had a cylindrical shape with a diameter of 48 mm and a length of 96 mm. The load was transferred from the hydraulic press to the sample by loading platens. Tensile strength could be calculated according to Equation (1) [30]:

$$T_s = \frac{2 \cdot L}{\pi \cdot d \cdot w} = 0.636 \cdot \frac{L}{d \cdot w} \quad (1)$$

where:

T_s—tensile strength (MPa);
L—maximal recorded load (N);
d—the diameter of the specimen (mm),
w—the width of the specimen (mm).

Sandstone at 300 °C and 600 °C increases its tensile strength by 24% and 10.6%, respectively, compared to the initial value. On the other hand, a further increase in temperature to the value of 900 °C and 1200 °C contributes to a decrease in strength by 55.36% and 78.3%, respectively. Even greater differences are found for wet samples. Only for the temperature of 300 °C is there an increase in strength by 19.6%, while for the temperatures of 600 °C, 900 °C and 1200 °C, there is a decrease by 5.37%, 72.8% and 93.45%, respectively. Immersion of the heated sandstone samples in water reduces the tensile strength by 4.4%, 15.97%, 17.44% and 20.15%, respectively, for the temperatures of 300 °C, 600 °C, 900 °C and 1200 °C compared to the only heated samples. The average decrease in tensile strength for claystone heated at 300 °C and treated with water was 55.15% and 75.78% with respect to the temperature of 20 °C. For samples of claystone heated at the temperature of 300 °C, the influence of water is characterized by an over two times decrease in tensile strength.

At the temperature of 300 °C and 600 °C, the value of Young's modulus increases by 3.96 GPa and 1.07 GPa, respectively, compared to the temperature of 20 °C. However, at temperatures of 900 °C and 1200 °C, the value drops and is lower by 5.1 GPa and 8.58 GPa, respectively, compared to the initial value. As a result of the action of water, the Young's modulus decreases by 0.79, 0.75, 1.58 and 1.4 GPa, respectively, for samples only heated at the temperatures of 300 °C, 600 °C, 900 °C and 1200 °C. In the case of claystone heated at 300 °C and when wet, the Young's modulus decreases by 3.89 GPa and 5.44 GPa, respectively. The action of water results in an almost six times lower value of Young's modulus compared to samples only heated for the maximum temperature of claystone.

For sandstone samples heated at 900 °C and 1200 °C, the Poisson's ratio value increases by 6% and 17.8%, respectively, and decreases by 18.8% and 10.9% for the temperatures of 300 °C and 600 °C in relation to the value of 20 °C. On the other hand, for the heated samples exposed to water, the value of the Poisson's ratio for the temperatures of 300 °C and 600 °C decreases by 24.3% and 15.1%, and for the temperatures of 900 °C and 1200 °C, the value increases by 1.1% and 23.6%, respectively. The influence of water on the heated samples reduces the Poisson's ratio by 8.16%, 6.04% and 5.9%, respectively, for the temperatures of 300 °C, 600 °C and 900 °C, and causes an increase of 3.5% for the temperature of 1200 °C. For claystone heated at 300 °C and wet claystone, the value of the Poisson's ratio increases to 13.1% and to 8.79% in relation to the temperature of 20 °C. As a result of the action of water, the value of the Poisson's ratio drops slightly by 2.29%. The summary of the compressive strength and Poisson's ratio in relation to high temperatures is presented in Table 1.

Table 1. Comparison of the results of compressive strength and Poisson's ratio in relation to high- temperatures.

Type of Rock	Parameter	Temperature, t (°C)				
		20	300	600	900	1200
Sandstone dry	Cs, (MPa)	61.2	110.6	83.6	42.8	21
	ν	0.241	0.196	0.215	0.256	0.284
Sandstone wet	Cs, (MPa)	59.6	103	64.2	23.4	2.54
	ν	0.238	0.18	0.202	0.240	0.294
Claystone dry	Cs, (MPa)	31.8	6.2			
	ν	0.234	0.27			
Claystone wet	Cs, (MPa)	26.8	1.78			
	ν	0.240	0.263			

Cs—compressive strength; ν—Poisson's ratio.

Heating sandstone and claystone at the temperature of 300 °C does not significantly affect the loss of weight. The losses are very small and amount to 0.3% and 0.24%. At the temperatures of 600 °C, 900 °C and 1200 °C, the losses are one order higher compared to the temperature of 300 °C. The loss of mass at these temperatures for claystone in relation to sandstone is greater by 1.24%, 2.53% and 2.94%, respectively.

With increasing temperature, the density of both sandstone and claystone decreases. However, much greater changes occur for claystone samples. For both rocks heated at the temperature of 300 °C, the change in density is at a similar level and amounts to about 1%. On the other hand, at the temperatures of 600 °C, 900 °C and 1200 °C, there is a further reduction in density, with the changes being greater for claystone by 1.98%, 3.35% and 9.54% compared to the sandstone samples.

4. Numerical Modeling

The main goal of numerical modeling was to determine the extent of the rock destruction zone around the gasification channel. For this purpose, RS2 [31] software was used, which is based on the finite element method. For the evaluation of the damage zones, the strength factor was selected, expressing the ratio of the rock strength to the reduced stresses at a given point. Strength factor values below 1 indicate material failure. The modeling adopted the Hoek–Brown criterion, which links the compressive strength and material constants determined using RocData software [31]. The results obtained from the laboratory tests were used in numerical simulations. A square target with a side length of 100 m was adopted in the modeling. In numerical modeling, it was assumed that the horizontal and vertical stresses are equal to each other: $\sigma_1 = \sigma_3 = \sigma_z = 23.3$ MPa. The adoption of such a value resulted from the depth of the hard coal deposit and the unit weight of the overburdened rocks. The size of the model was selected so that for the largest width of the gasification channel, equal to 30 m, the extent of rock destruction could be recorded. The models adopted a graded mesh and six-noded triangles with a gradation factor equal to 0.1. The number of nodes on all excavations was equal to 110. The models were restrained on all sides. As most of the claystone samples were destroyed under the influence of the temperature of 300 °C (Figure 3), the modeling assumed that the height of the gasification channel was 1.5 m. This height consisted of the thickness of the coal seam equal to 0.4 m and the total destruction of claystone rocks in the roof and the floor with a total thickness of 1.1 m (Figure 1). In the numerical simulations, three widths of the gasification channel, 10, 20 and 30 m (Figure 11a–c), were adopted. In addition, three ranges, 0.5, 1.0 and 1.5 m (Figure 11d–f), for the effects of high temperatures both in the roof and floor of the coal seam were modeled. The test results are shown in Figures 12–17 and Table 2.

Table 2. Summary of the results of the range of the rock destruction zone around the gasification channel.

Channel Width, (m)	Temperature Impact Range (m)				The Maximum Extent of the Rock Destruction Zone around the Gasification Channel for a Dry Rock Mass (m)		The Maximum Extent of the Rock Destruction Zone around the Gasification Channel for a Wet Rock Mass (m)	
	1200 °C	900 °C	600 °C	300 °C	Roof	Floor	Roof	Floor
10	0.5	1.0	1.5	2.0	4.51	4.36	5.0	4.9
	1.0	2.0	3.0	4.0	4.58	4.47	5.18	5.08
	1.5	3.0	4.5	6.0	4.63	4.54	5.52	5.32
20	0.5	1.0	1.5	2.0	8.96	8.60	9.75	9.23
	1.0	2.0	3.0	4.0	9.33	8.87	10.58	9.96
	1.5	3.0	4.5	6.0	9.49	9.04	10.92	10.30
30	0.5	1.0	1.5	2.0	14.06	12.91	15.27	13.98
	1.0	2.0	3.0	4.0	14.29	13.43	16.09	15.08
	1.5	3.0	4.5	6.0	14.72	13.70	16.83	15.58

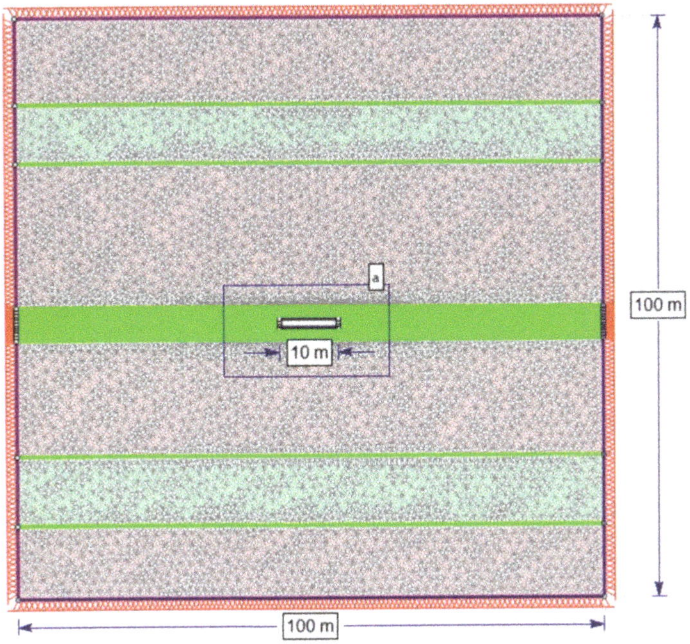

(a)

Figure 11. *Cont.*

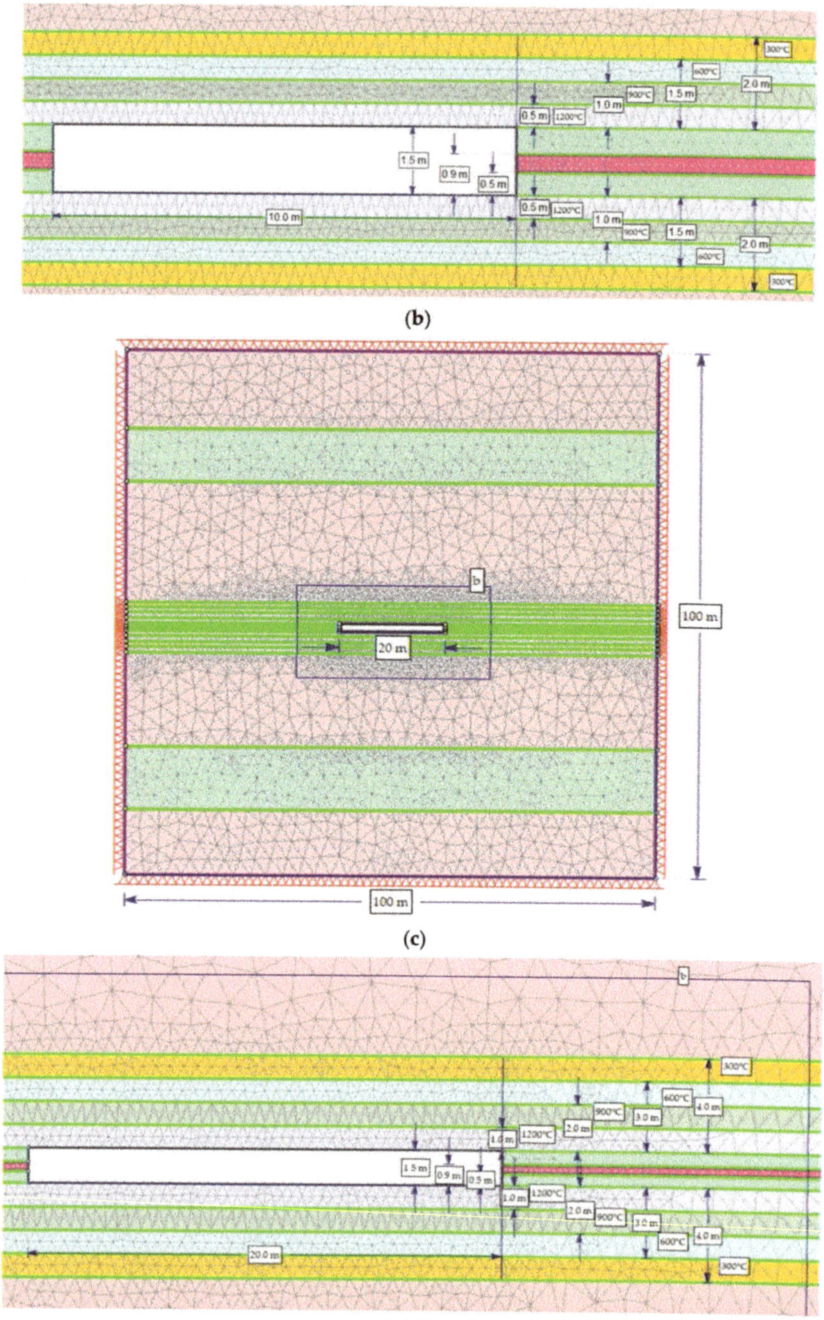

(b)

(c)

(d)

Figure 11. *Cont.*

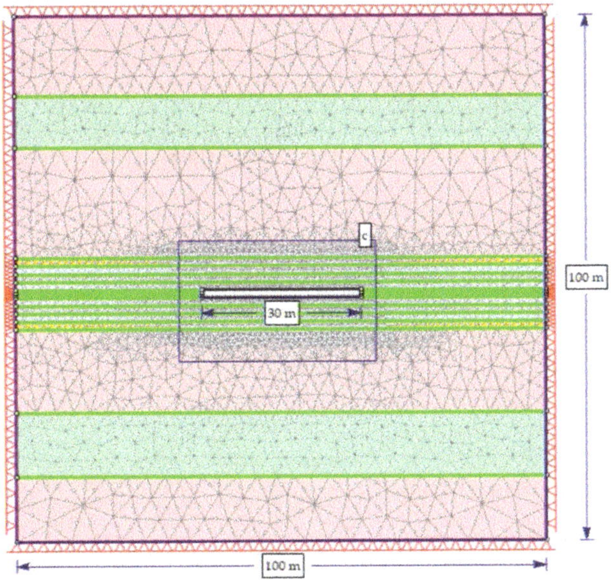

(e)

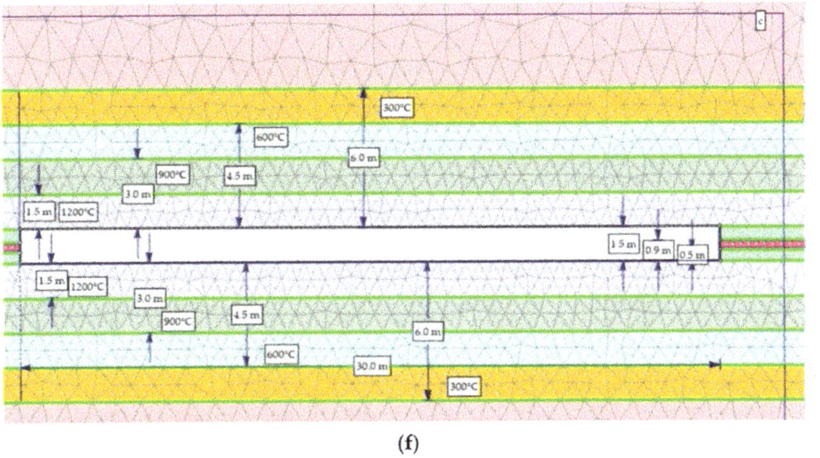

(**f**)

Figure 11. Numerical models: (**a**) channel width 10 m; (**b**) detail "a"; (**c**) channel width 20 m; (**d**); detail "b"; (**e**) channel width 30 m; (**f**) detail "c".

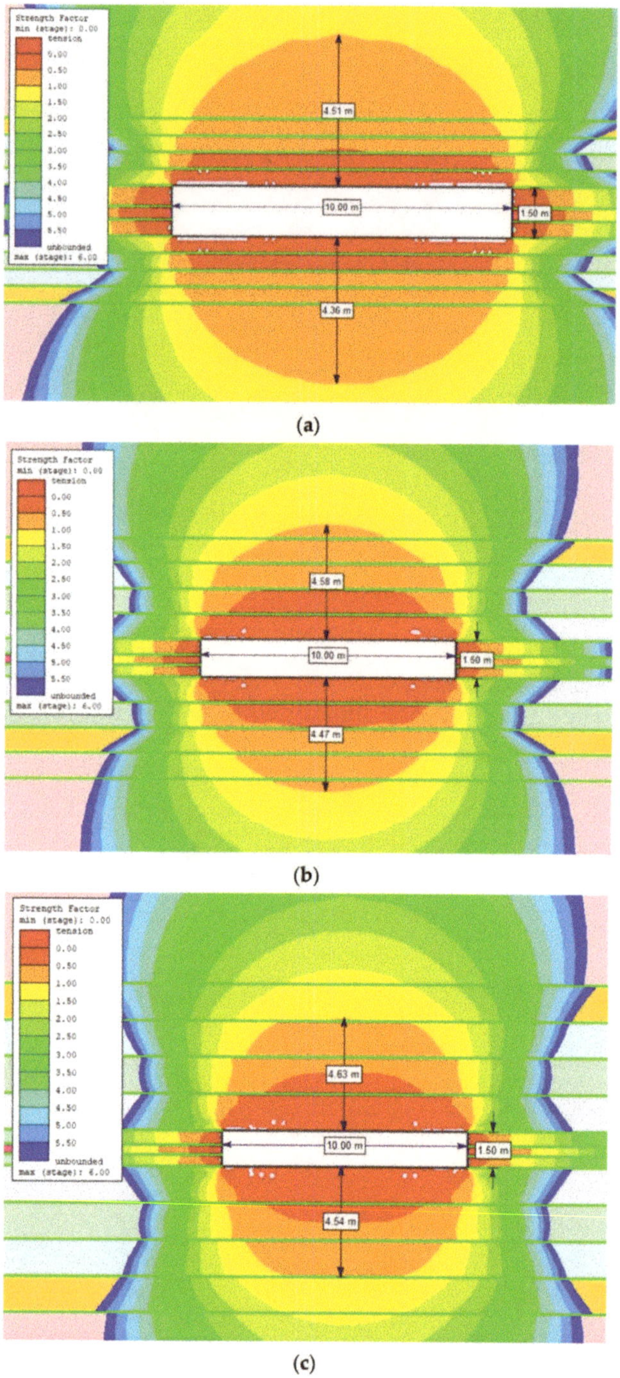

Figure 12. The extent of the rock destruction zone around the gasification channel with a width of 10 m and the impact of temperature of 1200 °C for the dry rock mass, up to the height of: (**a**) 0.5 m; (**b**) 1.0 m; (**c**) 1.5 m.

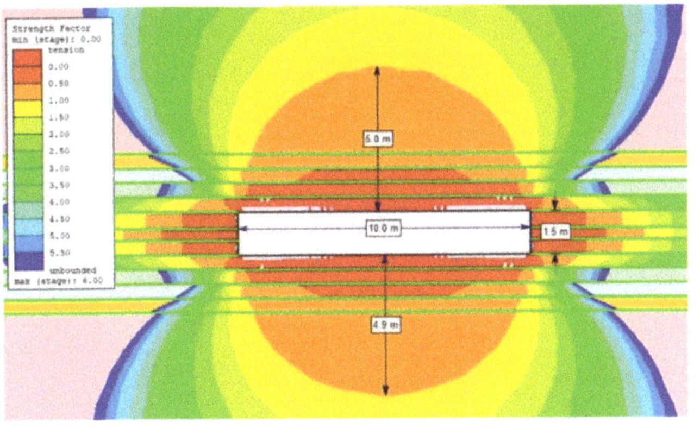

(**a**)

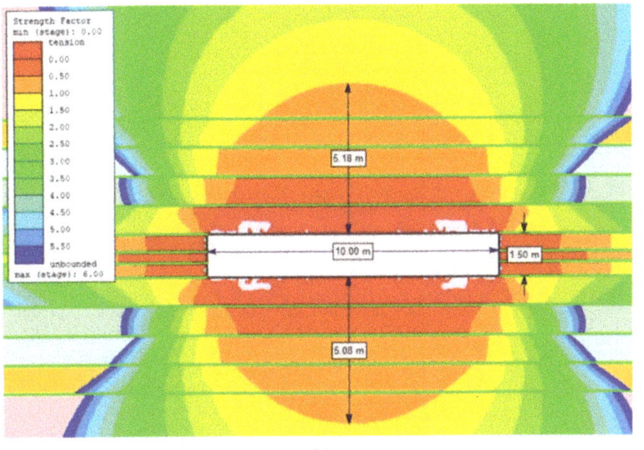

(**b**)

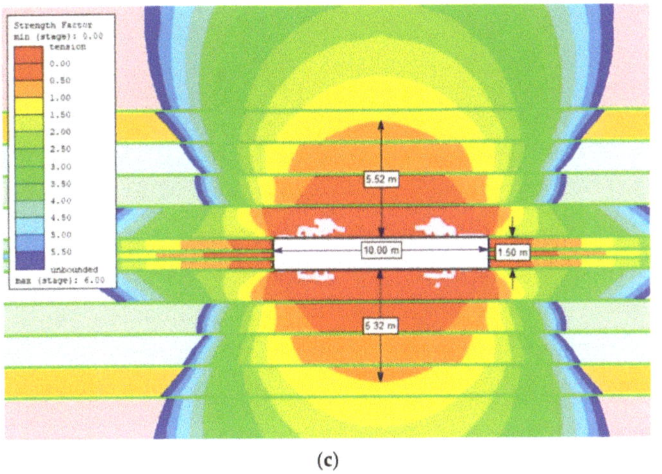

(**c**)

Figure 13. The extent of the rock destruction zone around the gasification channel with a width of 10 m and the impact of temperature of 1200 °C for the wet rock mass, up to the height of: (**a**) 0.5 m; (**b**) 1.0 m; (**c**) 1.5 m.

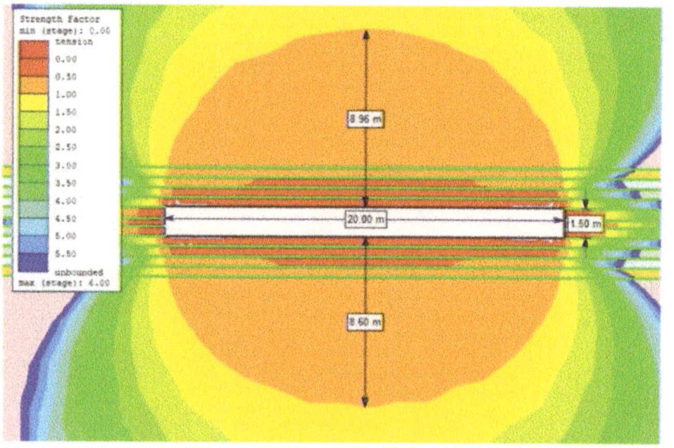

(a)

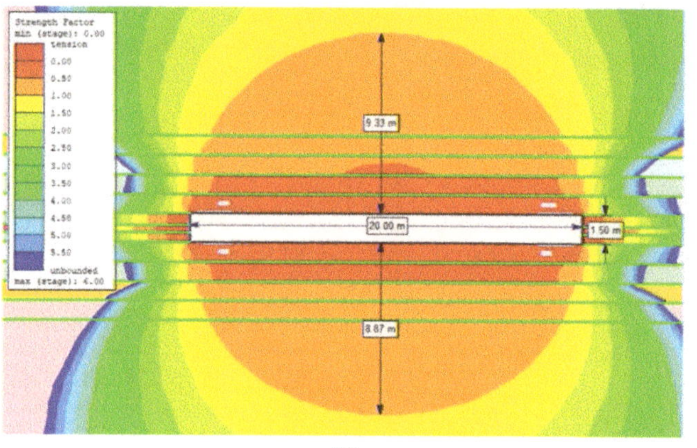

(b)

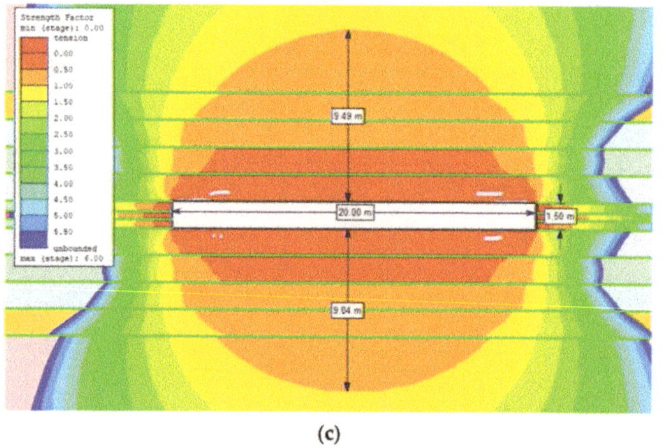

(c)

Figure 14. The extent of the rock destruction zone around the gasification channel with a width of 20 m and the impact of temperature of 1200 °C for the dry rock mass, up to the height of: (**a**) 0.5 m; (**b**) 1.0 m; (**c**) 1.5 m.

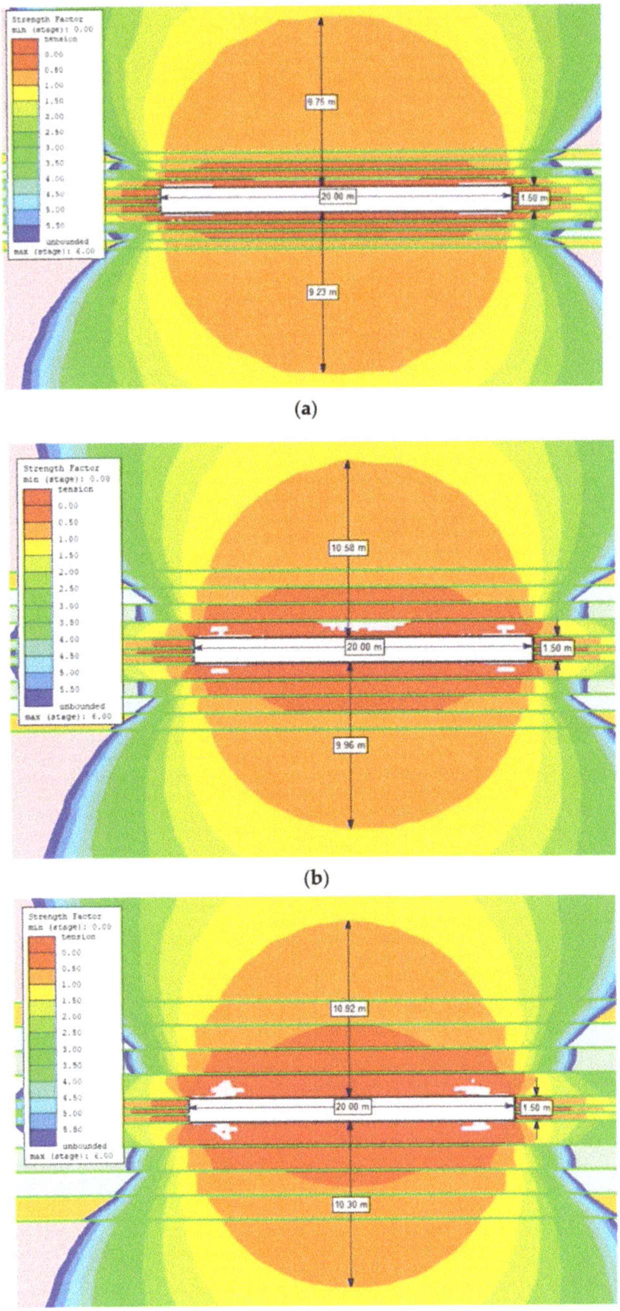

Figure 15. The extent of the rock destruction zone around the gasification channel with a width of 20 m and the impact of temperature of 1200 °C for the wet rock mass, up to the height of: (**a**) 0.5 m; (**b**) 1.0 m; (**c**) 1.5 m.

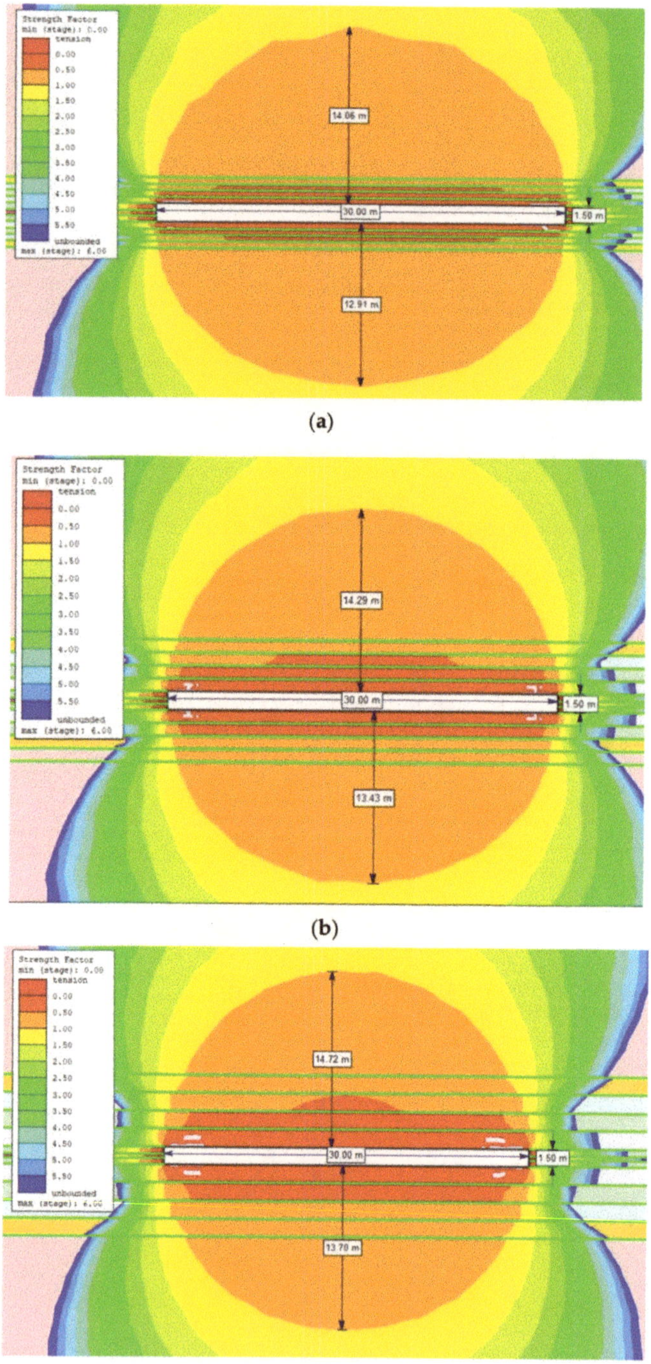

Figure 16. The extent of the rock destruction zone around the gasification channel with a width of 30 m and the impact of temperature of 1200 °C for the dry rock mass, up to the height of: (**a**) 0.5 m; (**b**) 1.0 m; (**c**) 1.5 m.

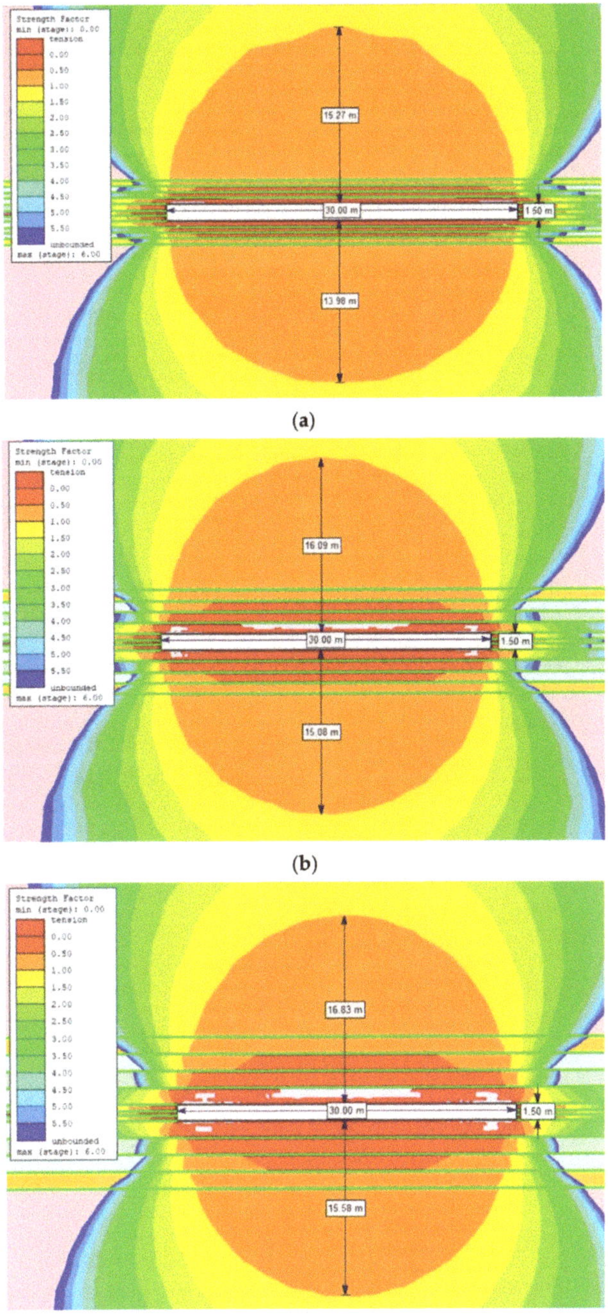

Figure 17. The extent of the rock destruction zone around the gasification channel with a width of 30 m and the impact of temperature of 1200 °C for the wet rock mass, up to the height of: (**a**) 0.5 m; (**b**) 1.0 m; (**c**) 1.5 m.

5. Discussion

The process of underground coal gasification takes place in underground generators, which can be prepared using shaft and non-shaft methods. In the shaft method, mining excavations are performed underground after the coal seam is made available through a vertical and inclined shaft. In the structure of accessibility, the workings of the mine are used to supply an oxidizing agent and discharge gas to the surface. The oxidizing agent may be supplied to the coal seam by means of: roadway inside the body of coal; through holes made in the body of coal between roadways; and blind holes to which the oxidizing agent is led through heat-resistant pipes [32]. The main advantage of this method is the accuracy of recognizing the conditions of the coal seam, easy drainage of the deposit and verification of the influence of high temperatures on the surrounding rock layers. On the other hand, the disadvantage is the cost of construction and the maintenance of access and preparatory workings. In the non-shaft method, the coal seam is made available through vertical, inclined and directional holes from the surface, and then they are joined together [33]. Vertical holes are used in the execution of the initial front line of fire and drainage of generators. Inclined holes are used when it is necessary to arrange them outside the zone of rocks with low strength parameters. Directional holes are a combination of access and generator holes, which are made in horizontal locations or at a slight angle.

Macroscopic observations of carboniferous rock samples taken from a closed hard coal mine in the Upper Silesian Coal Basin, heated at temperatures up to 1200 °C and exposed to water, show that sandstones are not subject to decomposition. Tian et al. [34] noticed that sandstone heated at 1000 °C does not lose its cohesion. Claystone samples, when heated to a temperature of 300 °C, burn out and break into small pieces. Wolf et al. [35] found that for claystone samples, the weight loss is due to, inter alia, the dehydroxylation of clays. Hetema et al. [36] determined that for claystone samples, the compaction number increased with increasing temperature. The influence of water on the heated sandstone samples at a temperature of 1200 °C contributes to the reduction in strength, deformation and structural parameters by 87% (compressive strength), 70% (tensile strength), 73% (Young's modulus) and 5.3% (density). The value of the Poisson's ratio is at a similar level; at 1200 °C, it slightly increases by 3.5%. Wu et al. [37] also confirmed the increase in Poisson's ratio after exceeding the temperature of 1000 °C. The behavior of rocks under the influence of heat and water is very diverse. Bresser et al. [38] discovered that for marbles heated at 600 °C, the influence of water on the change of strength parameters is small. Luo et al. [39] noted that the value of modulus of elasticity and compressive strength for mudstone at 200 °C ÷ 600 °C is at a similar level, followed by their increase. Kiliç [40] pointed out that for limestone samples, the weight loss process starts at 600 °C, whereas for sandstone and claystone samples, the weight loss is 2.6% and 3.9%, respectively. Zhang et al. [41] found that sandstone increases its compressive strength to a temperature of 500 ÷ 600 °C, followed by a drop in strength by several dozen percent. In turn, Rao et al. [42] noticed that sandstone increases its tensile strength to a temperature of 250 °C. The increase in compressive strength of sandstone samples is related to the mineralogical composition. The main component of sandstone is quartz, for which the melting point is above 1410 °C. At temperatures up to 300 °C, the dehydroxylation of iron oxides and hydroxides or organic matter oxidation can occur. On the other hand, up to the temperature of 600 °C, water vapor is released in the amount of several grams per kilogram of sample. The main mineralogical changes in sandstones as a result of high temperatures are the appearance of hematite and ore minerals and a reduction in the amount of heavy minerals [43]. The tensile strength, determined by the Brazilian method, is significantly influenced by the cross-sectional area along which the fracture occurs. This influence is conditioned by both surface and volumetric factors, with the volumetric effect being of decisive importance. It depends on the mineral composition, structure and texture of the rock and, above all, the sum of structural defects, in particular fracture and cleavage. Thus, the volume factor expresses the sum of the structural defects of the rock within a given volume, while the surface factor determines the condition of the potential fracture surface. For claystone rocks, both of the described factors appeared at the temperature of 300 °C, which resulted in a significant reduction in tensile strength. For sandstone rocks, the influence of the volumetric and surface factors was revealed only at the temperature of 900 °C. Undoubtedly, the change of strength parameters is significantly influenced by carbon substance lamines, which, when exposed to high temperature, burn out and directly reduce the tensile strength, especially in claystone rocks. The dependence of the compressive strength and tensile strength in relation to the decrease in density is shown in Figure 18a,b.

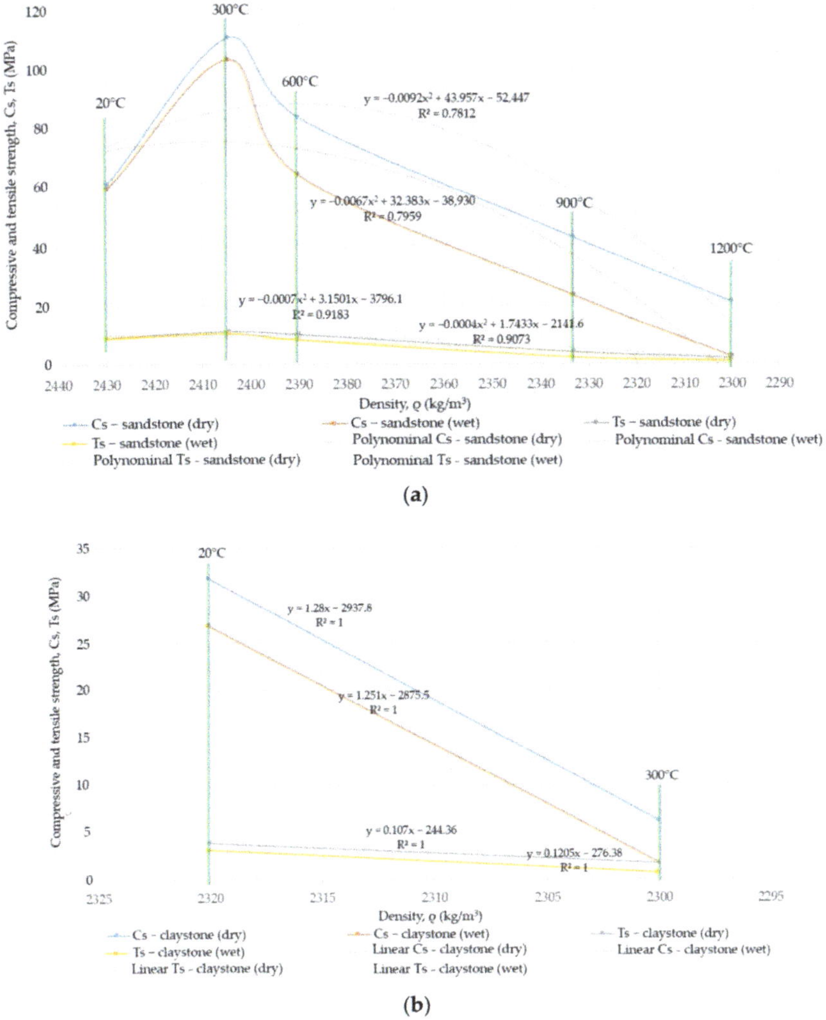

Figure 18. The dependence of the compressive strength and tensile strength in relation to the decrease in density for: (**a**) sandstone; (**b**) claystone.

For sandstone rocks heated at temperatures from 300 °C to 600 °C, the strength parameters increased by 80.7% and 36.6%, respectively (compressive strength for dry rocks); by 72.8% and 7.7%, respectively (compressive strength for wet rocks), by 24.06% and 10.06%, respectively (tensile strength for dry rocks); an increase of 19.63% and a decrease of 5.37% (tensile strength for wet rocks), accompanied by a decrease in density by 1.03% and 1.62%, respectively. For the temperatures of 900 °C to 1200 °C, the compressive strength decreased by 30.06% and 65.68% (for dry rocks) and 60.73% and 95.73% (for wet rocks), respectively. In the case of tensile strength, also for these temperatures, there was a decrease by 55.36% and 78.32 (for dry rocks) and 72.86% and 87.29% (for wet rocks), respectively, which was accompanied by a decrease in density by 3.97% and 5.33%, respectively. For claystone rocks heated to the temperature of 300 °C, both in the case of compressive strength and tensile strength, the value decreased by 80.5% (compressive strength dry) and 93.35% (compressive strength dry) and 55.15% (tensile strength dry) and 75.78% (tensile strength wet), accompanied by a decrease in density by 0.86%. A further increase in temperature to the values of 600 °C, 900 °C and 1200 °C contributed to a decrease in density by 3.62%, 7.32% and 14.87%, respectively, with the simultaneous destruction of the integrity of the samples, making it impossible to determine the strength parameters.

In the model studies conducted by Otto et al. [44] and Nakaten et al. [45], the width of the gasification channel was often assumed to be 20 m. Pivnyak et al. [46] determined that the range of the impact of high temperature in the roof and floor can reach values up to 5.7 and 3.2 thickness of the gasified coal seam, respectively. Experiments conducted in Russia showed that, depending on the type of rock, due to the heating of the roof, rocks in the cave zone may be in the range of 1.33–3.8 m, and the metamorphized zone of rocks is from 0.65 m to 0.84 m [47]. Luo et al. [48], based on numerical simulations, determined that the temperature of 900 °C can have a range of up to 9 m in the roof. On the other hand, Wiatowski et al. 2021 [49], based on research on large samples of hard coal, found that when there is a siderite layer in the coal seam, the maximum temperature of 1200 °C occurs at a distance of 0.3 m above the gasified seam. As a result of the high temperature, the roof rocks, in particular claystone, can fall downwards, making it possible for gases to migrate into the rock mass and cause an increase in the temperature of the surrounding rocks. Moreover, as a result of high temperature, the geometry of the carbon pillars changes, which translates into an increase and change in the stress distribution around the gasification channel [50].

6. Conclusions

Based on the laboratory tests of rock samples heated at high temperatures, it can be concluded that:

- For dry claystone at the temperature of 300 °C, the compressive strength and tensile strength decrease by 80.5% and 55.15%, respectively, while for wet claystone, these decrease by 93.35% and 75.78%, respectively, in relation to the initial value. After exceeding the temperature of 300 °C, the claystone decomposed;
- For dry sandstone heated at 300 °C and 600 °C, the compressive strength increases by 80.7% and 36.6%, respectively, while for temperatures from 900 °C to 1200 °C, there is a decrease by 30.06% and 65.68%, respectively, in relation to the temperature of 20 °C. In the case of wet sandstone, there is an increase of 72.8% and 7.7% for temperatures of 300 °C and 600 °C, respectively, and a decrease of 60.73% and 95.73%, respectively, for temperatures of 900 °C and 1200 °C. The tensile strength for dry sandstone increases by 24.06% and 10.06%, respectively, for the temperatures of 300 °C and 600 °C, and decreases by 30.06% and 65.68%, respectively, for the temperatures of 900 °C and 1200 °C. On the other hand, for wet sandstone, there is an increase of 19.63% for the temperature of 300 °C and a decrease of 5.37%, 72.86% and 87.29% for the temperatures of 600 °C, 900 °C and 1200 °C, respectively;
- Within the temperature range of 300 °C, 600 °C, 900 °C, 1200 °C, the density decreases by 0.86%, 3.62%, 7.32% and 14.87%, respectively, for claystone rocks, and 1.03%, 1.62%, 3.97% and 5.33%, respectively, for sandstone compared to the initial value.
- Based on numerical research, it can be concluded that:
- For the width of the gasification channel equal to 10, 20 and 30 m, the maximum extent of rock destruction for dry rock mass does not exceed 5, 10 and 15 m, respectively;
- An increase in the extent of rock destruction occurs for wet rock mass. For roof rocks, the maximum range is increased by 19.2%, 15% and 14.33%;

- Additionally, for floor rocks, there is an increase by 17.1%, 13.9% and 13.7% in relation to the dry rock mass.

Rocks located in the immediate vicinity of the designed georeactor are the basis for its stability analyses. The conducted research shows a strong dependence of certain values of the strength parameters of carboniferous rocks on high temperature, which in the process of underground coal gasification often exceeds 900 °C. The obtained results regarding the extent of the rock destruction zone around the gasification channel indicate the need to conduct this process in a strongly controlled manner, because the extent of rock destruction increases with the increase in the width of the gasification channel. The existence of underground water reservoirs and the possibility for water to tear through the fractured rock mass additionally reduces the strength parameters of the heated rocks and increases the extent of rock destruction.

Author Contributions: Conceptualization, K.S. and K.Z.; methodology, K.S. and K.Z.; software, K.S.; validation, K.S., K.Z. and A.Z.; formal analysis, K.S., K.Z. and A.Z.; investigation, K.S., K.Z. and A.Z.; resources, K.S.; data curation, K.S., K.Z. and A.Z.; writing—original draft preparation, K.S. and K.Z.; writing—review and editing, K.S., K.Z. and A.Z.; visualization, K.S., K.Z. and A.Z.; supervision, K.S., K.Z.; project administration, K.S., K.Z.; funding acquisition, KS., K.Z. All authors have read and agreed to the published version of the manuscript.

Funding: This research was prepared as part of AGH University of Science and Technology in Poland, scientific subsidy under number: 16.16.100.215.

Institutional Review Board Statement: Not applicable.

Informed Consent Statement: Not applicable.

Data Availability Statement: The data presented in this study are new and have not been previously published.

Conflicts of Interest: The authors declare no conflict of interest.

References

1. Kostúr, K.; Laciak, M.; Durdan, M. Some Influences of Underground Coal Gasification on the Environment. *Sustainability* **2018**, *10*, 1512. [CrossRef]
2. Dubiński, J.; Turek, M. Basic aspects of productivity of underground coal gasification proces. *Archi. Min. Sci.* **2015**, *2*, 443–453.
3. Sajjad, M.; Rasul, M.G. Prospect of Underground Coal Gasification in Bangladesh. *Procedia Eng.* **2015**, *105*, 537–548. [CrossRef]
4. Ordinance of the Minister of the Environment of June 20, 2005 Changing the Ordinance on the Criteria for the Balance of Mineral Deposits. Available online: https://www.prawo.pl/akty/dz-u-2005-116-978,17199479 (accessed on 26 August 2021).
5. Kapusta, K.; Stańczyk, K. Pollution of water during underground coal gasification of hard coal and lignite. *Fuel* **2011**, *90*, 1927–1934. [CrossRef]
6. Kapusta, K.; Stańczyk, K.; Korczak, K.; Pankiewicz, M.; Wiatowski, M. Some aspects of impact of underground coal gasification process on water environment. *Res. Rep. Min. Environ.* **2010**, *4*, 17–27. (In Polish)
7. Liu, S.; Wang, Y.; Yu, L.; Oakey, J. Volatilization of mercury, arsenic and selenium during underground coal gasification. *Fuel* **2006**, *85*, 1550–1558. [CrossRef]
8. Chmura, K. *Physico-Thermal Properties of Rocks of Some Polish Mining Basins*; Śląsk Publishing House: Katowice, Poland, 1970; p. 63. (In Polish)
9. Tian, H.; Mei, G.; Jiang, G.S.; Qin, Y. High-Temperature Influence on Mechanical Properties of Diorite. *Rock Mech. Rock Eng.* **2017**, *50*, 1661–1666. [CrossRef]
10. Liu, X.; Guo, G.; Li, H. Study on the propagation law of temperature field in surrounding rock of underground coal gasification (UCG) combustion cavity based on dynamic thermal parameters. *Results Phys.* **2019**, *12*, 1956–1963. [CrossRef]
11. Perkins, G. Underground coal gasification Part II: Fundamental phenomena and modeling. *Prog. Energy Combust. Sci.* **2018**, *67*, 234–274. [CrossRef]
12. Min, X.; Lin, X.; Weitao, L.; Xiangming, H.; Weimin, C.; Chao, L.; Zhigang, W. Study on the physical properties of coal pyrolysis in underground coal gasification channel. *Powder Technol.* **2020**, *376*, 573–592.
13. Feng, B.; Bhatia, S.K. Variation of the pore structure of coal chars during gasification. *Carbon* **2003**, *41*, 507–523. [CrossRef]
14. Deming, Z.; Wenwen, W.; Libo, L.; Hui, J.; Liejin, G. Variation of pore structure in Zhundong coal particle with stepped K_2CO_3 loading during supercritical water gasification. *Fuel* **2021**, *305*, 1–8.
15. Brotóns, V.; Tomás, R.; Ivorra, S.; Alarcón, J.C. Temperature influence on the physical and mechanical properties of a porous rock: San Julian's calcarentine. *Eng. Geol.* **2013**, *167*, 117–127. [CrossRef]

16. Yavuz, H.; Demirgas, S.; Caran, S. Thermal effect on the physical properties of cabonate rocks. *Int. J. Rock Mech. Min. Sci.* **2010**, *47*, 94–103. [CrossRef]
17. Chaki, S.; Takarli, M.; Agbodjan, W.P. Influence of thermal damage on physical properties on a granite rock: Porosity, permeability and ultrasonic wave evolutions. *Constr. Build Mater.* **2008**, *22*, 1456–1461. [CrossRef]
18. Tian, H.; Kempka, T.; Xu, N.X.; Ziegler, M. Physical properties of sandstone after high temperature treatment. *Rock. Mech. Rock. Eng.* **2012**, *45*, 1113–1117. [CrossRef]
19. Małkowski, P.; Niedbalski, Z.; Hydzik-Wiśniewska, J. The change and structural thermal properties of rocks exposed to high temperature in the vicinity of designed geo-reactor. *Archi. Min. Sci.* **2013**, *58*, 465–480.
20. Yang, L.; Liang, J.; Yu, L. Clean Coal technology—Study on the pilot Project experiments of underground coal gasifications. *Energy* **2003**, *28*, 1445–1460. [CrossRef]
21. Kacur, J.; Laciak, M.; Durdán, M.; Flegner, P. Model-Free Control of UCG Based on Continual Optimization of Operating Variables: An Experimental Study. *Energies* **2021**, *14*, 4323. [CrossRef]
22. Sygała, A.; Bukowska, M. Identification of temperature effect on post-critical geomechanical properties of loaded sandstones. *Arab. J. Geosci.* **2019**, *12*, 1–10. [CrossRef]
23. Otto, C.; Kempka, T. Thermo-Mechanical Simulations of Rock Behavior in Underground Coal Gasification Show Negligible Impact of Temperature-Dependent Parameters on Permeability Changes. *Energies* **2015**, *8*, 5800–5827. [CrossRef]
24. Hsu, C.; Davies, P.T.; Wagner, N.J.; Kauchali, S. Investigation of cavity formation in lump coal in the context of underground coal gasification. *J. South. Afr. Inst. Min. Metall.* **2014**, *114*, 305–309.
25. Zagorščaka, R.; Sadasivama, S.; Thomas, H.R.; Stańczyk, K.; Kapusta, K. Experimental study of underground coal gasification (UCG) of a high-rank coal using atmospheric and high-pressure conditions in an ex-situ reactor. *Fuel* **2020**, *270*, 117490. [CrossRef]
26. Yang, L.; Zhang, X.; Lis, S.; Yu, L.; Zhang, W. Field test of large-scale hydrogen manufacturing from underground coal gasification (UCG). *Int. J. Hydrog.* **2008**, *33*, 1275–1285. [CrossRef]
27. Falsztinskij, W.S.; Diczkowskij, R.E.; Łoziskij, W.G. Economical justification of effectiveness the sealing rockmass above the gas generator for borehole coal gasification. *Res. Rep. Min. Environ.* **2010**, *3*, 51–59. (In Polish)
28. Polish Standard: PN-EN 1936. *Natural Stone Research Methods—Determination of Density and Bulk Density as well as Total and Open Porosity*; Publishing Polish Committee for Standardization: Warszawa, Poland, 2010; pp. 1–12. (In Polish)
29. International Society for Rock Mechanics and Rock Engineering. Suggested Methods for Determining Tensile Strength of Rock Materials Part 2: Suggested Method for determining indirect tensile strength by the Brazil Test. *Int. J. Rock Mech. Min. Sci.* **1978**, *15*, 99–103. [CrossRef]
30. American Society of Testing and Materials. *Standard Test Method for Splitting Tensile Strength of Intact Rock Core Specimens*; ASTM D3967-16; ASTM International: West Conshohocken, PA, USA, 2016; pp. 1–5.
31. Rocscience. Available online: https://www.rocscience.com (accessed on 11 September 2021).
32. Białecka, B. Underground coal gasification. In *Basics of the Decision-Making Process*; Publishing House of the Central Mining Institute: Katowice, Poland, 2008; pp. 25–28. (In Polish)
33. Khan, M.M.; Mmbaga, J.P.; Shirazi, A.S.; Trivedi, J.; Liu, Q.; Gupta, R. Modelling Underground Coal Gasification—A Review. *Energies* **2015**, *8*, 12603–12668. [CrossRef]
34. Tian, H.; Kempka, T.; Schlüter, R.; Feinendegen, M.; Ziegler, M. Influence of high temperature on rock mass surrounding in situ coal conversion sites. In Proceedings of the 10th International Symposium on Environmental Geotechnology and Sustainable Development, Bochum, Germany, 7–11 September 2009; pp. 128–132.
35. Wolf, K.H.A.A.; Hettema, M.H.H.; Pater, C.J.; Van Hooydonk, R. Classification of overburden properties for underground coal gasification: Laboratory studies under high temperature and in situ stress conditions. In Proceedings of the Rock Characterization: ISRM Symposium, Eurock, Chester, UK, 14–17 September 1992; pp. 93–98.
36. Hettema, M.H.H.; Niepce, D.V.; Wolf, K.H.A.A. A microstructural analysis of the compaction of claystone aggregate at high temperature. *Int. J. Rock Mech. Min. Sci.* **1999**, *36*, 57–68. [CrossRef]
37. Wu, G.; Wang, Y.; Swift, G.; Chen, J. Laboratory investigation of the effect of temperature on the mechanical properties of sandstone. *Geotech. Geol. Eng.* **2013**, *31*, 809–816. [CrossRef]
38. Bresser, J.H.P.; Urai, J.L.; Olgaard, D.L. Effect of water on the strength and mictrostructure of Carrara marble axially compressed at high temperature. *J. Struct. Geol.* **2005**, *27*, 265–281. [CrossRef]
39. Luo, J.; Wang, L. High-temperature mechanical properties of mudstone in the process of underground coal gasification. *Rock Mech. Rock Eng.* **2011**, *44*, 749–754. [CrossRef]
40. Kiliç, Ö. The influence of high temperature on limestone P-wave velocity and Schmidt hammer strength. *Int. J. Rock Mech. Min. Sci.* **2006**, *43*, 980–986. [CrossRef]
41. Zhang, L.; Mao, X.; Lu, A. Experimental study on the mechanical properties of rocks at high temperature. *Sci. China Ser. E-Technol. Sci.* **2009**, *52*, 641–646. [CrossRef]
42. Rao, Q.-H.; Wang, Z.; Xie, H.-F.; Xie, Q. Experimental study of mechanical properties of sandstone at high temperature. *J. Cent. Univ. Technol.* **2007**, *14*, 478–483. [CrossRef]
43. Małkowski, P.; Skrzypkowski, K.; Bożęcki, P. Changes of rock behaviour under the influence of high temperatures in the vicinity of a georeactor. *Res. Reports. Min. Environ.* **2011**, *4/2*, 259–272. (In Polish)

44. Otto, C.; Kempka, T. Thermo-mechanical simulations confirm: Temperature-dependent mudrock properties are nice to have in far-field environmental assessments of underground coal gasification. *Energy Procedia* **2015**, *76*, 582–591. [CrossRef]
45. Nakaten, N.; Kempka, T. Techno-Economic Comparison of Onshore and Offshore Underground Coal Gasification End-Product Competitiveness. *Energies* **2019**, *12*, 3252. [CrossRef]
46. Pivnyak, G.; Falshtynskyi, V.; Dychkovskyi, R.; Saik, P.; Lozynskyi, V.; Cabana, E.; Koshka, O. Conditions of Suitability of Coal Seams for Underground Coal Gasification. *Key Eng. Mater.* **2020**, *844*, 38–48. [CrossRef]
47. Dengina, N.I.; Kazak, V.N.; Pristash, V.V. Changes in rock at high temepraure. *J. Min. Sci.* **1994**, *29*, 472–477. [CrossRef]
48. Luo, J.; Lianguo, W.; Furong, T.; Yan, H.; Lin, Z. Variation in the temperature field of rocks overlying a high-temperature cavity during underground coal gasification. *Min. Sci. Technol.* **2011**, *21*, 709–713. [CrossRef]
49. Wiatowski, M.; Kapusta, K.; Nowak, J.; Szyja, M.; Basa, W. An exsitu underground coal gasification experiment with a siderite interlayer: Course of the process, production gas, temperatures and energy efficiency. *Int. J. Coal. Sci. Technol* **2021**, *8*, 1–14. [CrossRef]
50. Xu, Y.; Li, H.; Guo, G.; Liu, X. Stability analysis of hyperbolic coal pillars with peeling and high temperature effects. *Energy Explor. Exploit.* **2020**, *38*, 1574–1588. [CrossRef]

energies

Article

Effect of Lignite Properties on Its Suitability for the Implementation of Underground Coal Gasification (UCG) in Selected Deposits

Krzysztof Kapusta

Central Mining Institute, Plac Gwarków 1, 40-166 Katowice, Poland; kkapusta@gig.eu; Tel.: +48-32-3246535

Abstract: Two experimental simulations of underground coal gasification (UCG) processes, using large bulk samples of lignites, were conducted in a surface laboratory setup. Two different lignite samples were used for the oxygen-blown experiments, i.e., "Velenje" meta-lignite (Slovenia) and "Oltenia" ortho-lignite (Romania). The average moisture content of the samples was 31.6wt.% and 45.6wt.% for the Velenje and Oltenia samples, respectively. The main aim of the study was to assess the suitability of the tested lignites for the underground coal gasification process. The gas composition and its production rates, as well as the temperatures in the artificial seams, were continuously monitored during the experiments. The average calorific value of gas produced during the Velenje lignite experiment (6.4 MJ/Nm^3) was much higher compared to the result obtained for the experiment with Oltenia lignite (4.8 MJ/Nm^3). The Velenje lignite test was also characterized by significantly higher energy efficiency, i.e., 44.6%, compared to the gasification of Oltenia lignite (33.4%). The gasification experiments carried out showed that the physicochemical properties of the lignite used considerably affect the in situ gasification process. Research also indicates that UCG can be considered as a viable option for the extraction of lignite deposits; however, lignites with a lower moisture content and higher energy density are preferred, due to their much higher process efficiency.

Keywords: lignite; underground coal gasification; UCG; ex situ tests

Citation: Kapusta, K. Effect of Lignite Properties on Its Suitability for the Implementation of Underground Coal Gasification (UCG) in Selected Deposits. *Energies* **2021**, *14*, 5816. https://doi.org/10.3390/en14185816

Academic Editors: Marek Laciak, Ján Kačur and Milan Durdán

Received: 3 August 2021
Accepted: 9 September 2021
Published: 14 September 2021

Publisher's Note: MDPI stays neutral with regard to jurisdictional claims in published maps and institutional affiliations.

1. Introduction

Despite the current shift to renewable sources of energy, fossil fuels, and, in particular, coal, will remain a meaningful fuel in some parts of the world, for a long time into the future [1]. Increased energy demand will ultimately lead to the mining of deep coal deposits. As a consequence of the increasing depths, conventional underground mining is more difficult, more dangerous, and more expensive nowadays. Underground coal gasification is considered to be a technology that will enable the safe and economical exploitation of coal resources that could not otherwise be mined [2–4]. In recent decades, advances in directional drilling and monitoring technologies have changed the way that UCG can be conducted, at great depths. This, coupled with the current energy security issues and the need to reduce the environmental footprint, has sparked a global revival of interest in UCG. During UCG, coal deposits are transformed into gaseous combustible products (syngas) and extracted to the surface. The quality of the product gas depends, to a large extent, on the gasification medium used, operating conditions, coal rank, and local hydrogeology [5–10]. UCG can be a viable extraction technology for coal seams, for which conventional coal mining technologies are technically, economically, or environmentally not feasible [11]. Moreover, with the high-profile mine disasters, UCG is deemed safer than conventional mining, since it does not require staff below the ground.

Low-value coals, predominantly lignites, constitute a considerable proportion of global coal reserves [12]. The specific physicochemical properties of lignites result in many limitations in their extraction and further utilization. Lignites are characterized by high humidity, up to 60 wt.%, and thus low calorific values [13]. The suitability of lignites for UCG is therefore questionable [14]. The high moisture content of typical lignite results in low thermal efficiency and poor quality of gas from UCG, because a significant amount of thermal energy is used to evaporate the water. Another challenge would be performing UCG at relatively shallow depths, which may negatively affect the groundwater quality [15–17]. Although UCG has been tested in many parts of the world, many technological issues still need to be solved before it is commercially applied.

The former UCG tests using lignite samples showed that the moisture content is one of the key parameters influencing the gasification performance, the quality of the gaseous products, and, finally, the suitability of lignite for UCG [18–21]. Lignite gasification using oxygen-enriched air (OEA) resulted in a low gas calorific value, 4.18 MJ/Nm^3, for the experimentally optimized oxygen/air ratio [18]. It was experimentally proven that the consumption of heat for water evaporation leads to a very poor thermal efficiency (20%). Another study showed that lignite gasification under elevated pressure resulted in increased methane formation, with a lower H_2 and CO content in the UCG gas produced [20]. During a multiday UCG experimental simulation on lignite, conducted in a large-scale ex situ test facility, gas with an average calorific value of about 7.2 MJ/Nm^3 was produced, with a process energy efficiency 59%. The results suggested that the crucial issue for the improvement of process efficiency in lignite seams may be an appropriate geometry of the underground reactor. Nevertheless, it was revealed that extraction of the high-moisture lignites using oxygen-blown UCG may be a feasible option [14]. This paper presents the results of laboratory experiments on the suitability of two European lignites for UCG. Two multiday UCG experimental simulations were conducted in a laboratory ex situ test stand, dedicated for experiments with large coal samples. Most of the experimental studies presented in the previous works were carried out for relatively small seam geometries. In the presented research, the volumes of the tested lignite samples significantly exceeded those used in the earlier studies. Such geometries allow for a more accurate estimation of the feasibility of the gasification process of the examined lignite deposits.

2. Materials and Methods

2.1. UCG Experimental Installation

The test facility used for the ex situ UCG tests is presented in Figures 1 and 2. The installation enables the simulation of the UCG process in laboratory conditions. The main section of the test facility is a gasification chamber, where in situ geological conditions are simulated both in respect to coal seam and surrounding strata. The maximum length of the artificial coal seam is about 7 m.

Oxygen, steam and air, supplied individually or in mixtures may be used as gasification reagents. N_2 is used as a safety and inertizing agent. The gas produced is water scrubbed to lower the temperature to remove particulates and gasification tar. Subsequent step involves the separation of aerosols. Gas is combusted in a thermal combustion. Concentrations of the main components are determined using chromatographic (GC) technique. The temperatures during the gasification tests are measured by 14 sensors (Pt10Rh-Pt) installed directly in various zones of the gasification chamber.

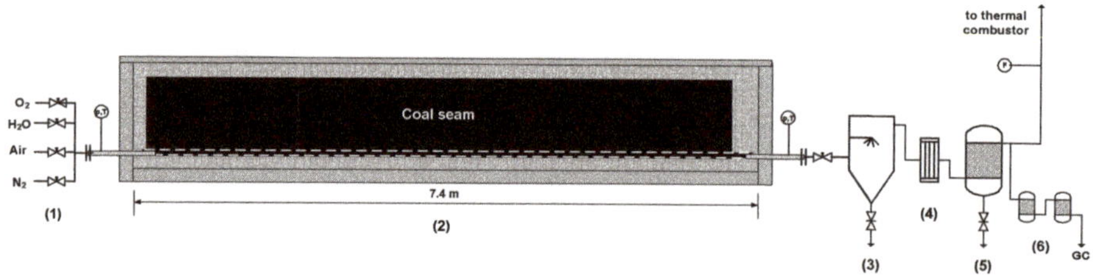

Figure 1. Scheme of the experimental installation: (1) supply of reagents, (2) gasification reactor, (3) scrubber, (4) cooler, (5) dust separator, (6) gas filters.

Figure 2. Ex situ UCG installation used in the study.

2.2. Lignite Samples and Creation the Coal Seams

The lignite samples for the gasification experiments were obtained from two different locations. The first selection of coal samples was gathered from an underground mine, Premogovnik Velenje, Slovenia. The sampling location was in the underground workings at a depth of approximately 350 m below the ground level (sample labelled as Velenje). The second selection of blocks was obtained from the Peşteana open cast mine (Complexul Energetic Oltenia Company), Romania. The sampling location was approximately 100 m below the ground level from a coal seam no. 8. This sample was labelled in the study as Oltenia. Physicochemical characteristics of the lignites used are showed in Table 1. Both the lignites are characterized by high moisture content. Oltenia sample is characterized by significantly higher moisture, ash and sulfur contents and lower content of volatile matter compared to lignite from Velenje. This resulted in considerably lower calorific value (as received basis). All analyses were performed by a certified laboratory (accreditation certificate according to ISO/IEC 17025).

The sulfur speciation studies revealed that in both lignites, pyritic sulfur is the main chemical form of sulfur (Figure 3). The Oltenia lignite is characterized by considerably higher content to total sulfur compared to the Velenje sample, i.e., 2.43% and 0.66%, respectively (analytical conditions). With respect to the UCG process, the ratio of ash sulfur-to-combustible sulfur is a crucial parameter governing partitioning of the total sulfur between post-gasification ash/slag left underground and product gas recovered on the surface. For the Oltenia lignite more than 90% of total sulfur occurs in combustible form. About 55% of total sulfur of Velenje lignite is supposed to remain as the post-gasification ash/slag.

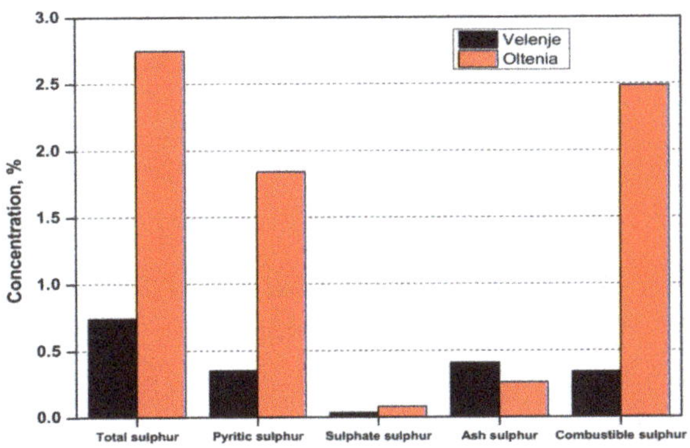

Figure 3. Sulfur forms in lignites used for ex situ gasification tests (dry basis).

Table 1. Physicochemical characteristics of lignites used for the ICG tests.

No.	Parameter	Lignite Sample	
		Velenje	Oltenia
	As received		
1	Total moisture W_t^r, %	31.62	45.64
2	Ash A_t^r, %	4.29	8.86
3	Volatiles V^r, %	43.67	25.78
4	Total sulfur S_t^r, %	0.51	1.49
5	Calorific value Q_i^r, kJ/kg	13,615	10,642
	Analytical		
6	Moisture W^a, %	11.13	11.49
7	Ash A^a, %	5.57	14.42
8	Volatiles V^a, %	56.76	41.98
9	Heat of combustion Q_s^a, kJ/kg	19,719	20,001
10	Calorific value Q_i^a, kJ/kg	18,427	18,860
11	Total sulfur S^a, %	0.66	2.43
12	Carbon C_t^a, %	49.86	49.49
13	Hydrogen H_t^a, %	4.67	3.94
14	Nitrogen N^a, %	0.64	1.34
15	Oxygen O_d^a, %	27.83	17.12

The ash fusion test describes the behavior of the ash residues at high temperature and it is a crucial factor in selecting solid fuels for energy use. The following ash parameters were determined for the two lignites under study:

- Sintering temperature;
- Softening temperature;
- Melting temperature;
- Flow (fluid) temperature.

The ash softening temperature is the temperature at which the ash softens beyond some arbitrary softness and the melting temperature is a measure of when the ash will melt and transform from solid to liquid. Since the UCG process has a zonal character (oxidation and reduction zones), the ash fusion temperatures are determined for both the oxidizing and reducing atmosphere. With regard to the UCG process, the key parameter is the ash flow temperature, at which slag begins to flow in the underground cavity/channels, which can cause clogging of the gas paths. Higher ash flow temperatures were observed for Velenje lignite, i.e., 1270 and 1300 °C, under oxidative and reducing conditions, respectively (Figure 4). Contrary to the Velenje sample, for Oltenia lignite, lower values of flow temperature were reported under reducing environment. This suggests that during UCG of Oltenia lignite, there may be a greater risk of channel clogging in the reduction (gasification) zone of the UCG reactor than in the oxidation zone. The observed significant differences in the sintering temperatures of both coal samples result from the different compositions of the mineral matter.

The raw samples provided by the industrial partners were used to prepare the continuous artificial coal seams of the total length of 6.0 m, width of 0.7 m and thickness of 0.7 m for experiments (Figure 5).

The gasification channel was prepared along the bottom part of the seams and its dimensions were 0.1 × 0.1 m. Sand was used to fill the voids between coal blocks and the reactor's walls as well as for the creation of the roof stratum. Fourteen temperature sensors were installed inside the gasification chamber to measure distributions of temperature during the UCG tests (Figure 5). The Nos. 1–4 were installed in the gasification channel, Nos. 5–14—inside the coal artificial seam.

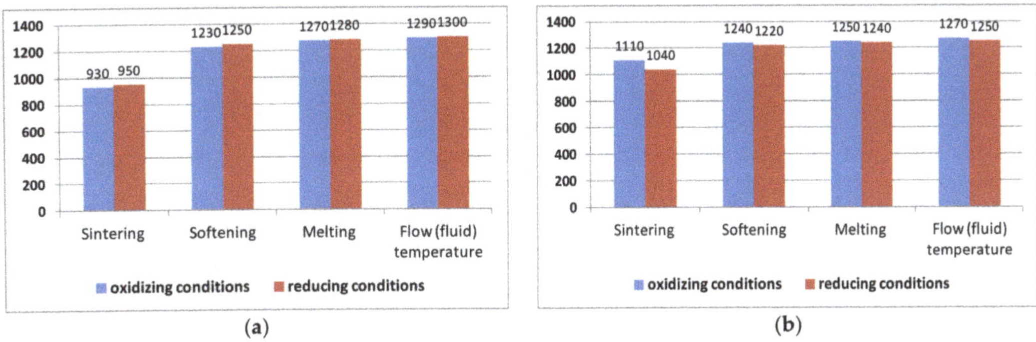

Figure 4. Ash fusion temperatures in oxidizing and reducing environments for lignites used for the UCG tests: (**a**) Velenje, (**b**) Oltenia.

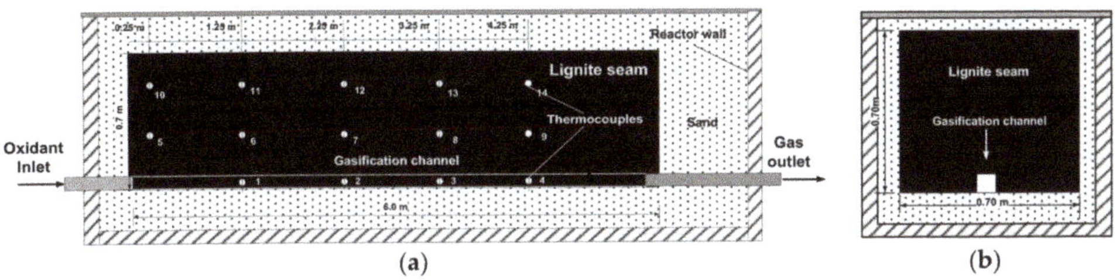

Figure 5. Sections of the artificial coal seam (proportions are not respected): (**a**) longitudinal (**b**) cross-section.

2.3. Gasification Test Procedure

The lignite samples were ignited using a pyrotechnic charge. The charge consisted of 800 g granulated pyrotechnic mass typically used in the mining industry. Appropriate modification of the composition allowed the achievement of an appropriately long combustion time (3–5 min) at high temperature (800–900 °C). The charge was placed 0.7 m from the front face of the seam. The pyrotechnic material was ignited by two fuses actuated by a capacitor electric igniter. The ignition was considered complete when the O_2 concentration in the outlet gas dropped below 1% vol. The gasification process was started by putting oxygen (99.5% purity) into the ignited coal seam. The initial oxygen supply rate was 2–3 m^3/h and it was gradually increased over the course of the experiment, up to a maximum value of 5 m^3/h in the final phases of the UCG tests. No additional water was supplied during the first test with Velenje lignite. In the second UCG trial with Oltenia lignite, a short test to investigate the influence of steam was carried out. The decision to add steam was made during the gasification experiment, taking into account the deteriorating quality of the gas produced and a rapid drop in the gas production rate (see Figure 6). Both UCG experiments were carried out under near-atmospheric pressure conditions. Concentrations of the gas components (H_2, CO, CH_4, CO_2, N_2, O_2, C_2H_6, H_2S) were analyzed in one hour time intervals. The UCG experiments lasted for 120 and 96 h for Velenje and Oltenia lignites, respectively.

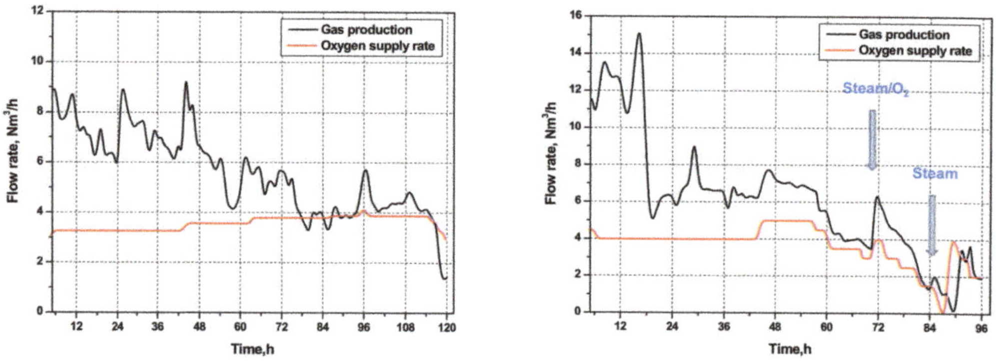

(a) (b)

Figure 6. Gas production rates and oxygen supply rates during the UCG experiments: (**a**) Velenje lignite, (**b**) Oltenia lignite.

3. Results and Discussion

3.1. Gas Production Rate and Gas Composition

The evolution of UCG gas product during the experiments, and the oxidant supply rates (Nm^3/h) are presented in Figure 6. As can be observed from the graphs, the values of the production rates changed over the course of the experiments, with maximum values of about 9 Nm^3/h and 15 Nm^3/h for the Velenje and Oltenia experiments, respectively. The relatively intensive gas production rates in the initial gasification stages were mostly because of an intensive coal devolatilization (pyrolysis) at the beginning of the process. This intense gas evolution was especially evident in the Oltenia experiment. This can be explained by the specific physicochemical properties of the lignite sample. The Oltenia sample is classified as soft a ortho-lignite, which, compared to the Velenje meta-lignite, is geologically younger, contains more moisture, and has lower mechanical strength. Consequently, as a result of the intensive heating at the early stage of the experiment, the gasification and pyrolysis reactions were more intense compared to the Velenje experiment, mainly due to the mechanical disintegration of the seam and the rapid release of volatiles. This resulted in the relatively high gas production rates.

Changes in the UCG gas composition and in the gas calorific value for the experiments carried out are presented in Figures 7 and 8, respectively. As can be observed from the graphs, the initial gasification periods for both the experiments were characterized by a good-quality product gas, with a relatively high calorific value. In both the experiments, from about the 48th hour, a gradual decrease in the gas production rate was observed, with a significant deterioration in the gas quality, expressed in the content of combustible components (H_2, CH_4, CO) and its calorific value. The very high concentrations of CO_2 observed in the gas suggest that the combustion was the main chemical reaction when the gasification conditions deteriorated. This may be due to a limited availability of water for the gasification process. Consequently, during the Oltenia test, in the 70th hour of the experiment, steam was additionally supplied to the reactor, at a constant rate of 2 kg H_2O/h. As can be observed from Figure 7b, this resulted in a slight increase in the gas quality, which then gradually decreased. Although the gas quality improved again after increasing the oxygen supply rate at the 84th hour of the experiment, the effect was short-lived. Since the addition of water did not significantly improve the calorific value of the gas, the deterioration in the gasification conditions could be associated with a change in cavern geometry and disturbances in the gas flow, resulting in the combustion of the gasification products.

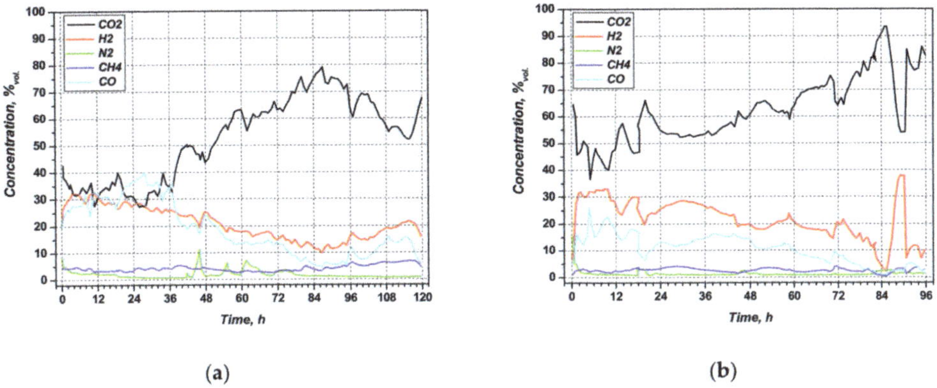

Figure 7. Changes in gas composition during the experiments: (**a**) Velenje lignite, (**b**) Oltenia lignite.

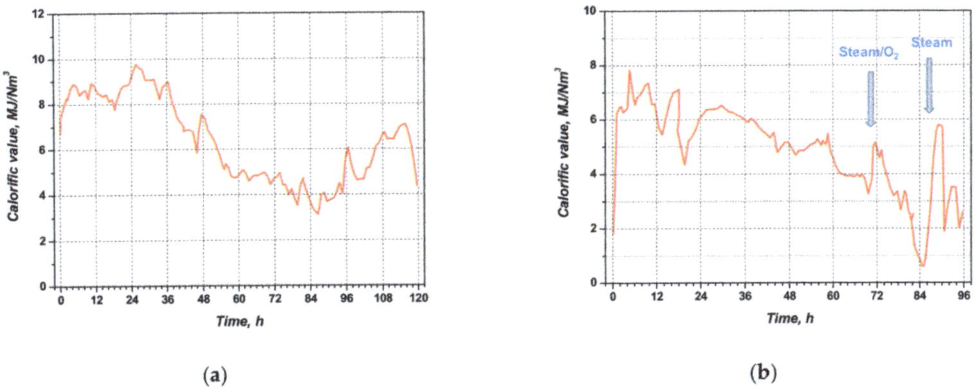

Figure 8. Changes in gas calorific value during the experiments: (**a**) Velenje lignite, (**b**) Oltenia lignite.

The average gas compositions for the experiments are presented in Table 2. The final gaseous products from both the experiments are characterized by relatively high hydrogen contents, i.e., 21.0% $vol_.$ and 21.3 % $vol_.$, for the experiments with Velenje and Oltenia lignite, respectively.

It was revealed that the Velenje gas product contained relatively high amounts of carbon monoxide compared to the product obtained during the gasification of Oltenia lignite. This may be due to the creation of more-favorable conditions for the Boudouard reaction, during the Velenje experiment, which resulted in a lower CO_2 concentration and higher calorific value of the gas. The average calorific value of gas, which was obtained for Velenje lignite - 6.4 MJ/Nm^3, was comparable with the results of a previous UCG experiment with Polish lignite, which resulted in gas with an average calorific value of 7.2 MJ/Nm^3 [14]. The UCG test was carried out in the same experimental installation and for a similar geometry of the artificial lignite seam. The differences in gas composition can be attributed to the coal properties, e.g., elemental composition, petrography, and ash mineralogy. The average calorific value of gas produced during the Oltenia lignite experiment (4.8 MJ/Nm^3) was much lower compared to the result obtained for the experiment with Velenje lignite. This is mainly due to the higher moisture and ash content in the Oltenia lignite sample used for the test (Table 1). Therefore, the conducted study showed that the physicochemical properties of lignite strongly affect the chemical composition of the UCG gas produced.

Table 2. Average compositions of gas obtained in the UCG tests.

Lignite Sample	Gas Composition, %vol.								Q, MJ/Nm^3
	CO_2	C_2H_6	H_2	O_2	N_2	CH_4	CO	H_2S	
Velenje	52.5	0.2	21.0	1.0	2.0	4.3	18.6	0.5	6.4
Oltenia	63.3	0.2	21.3	0.2	1.5	2.7	10.2	0.6	4.8

3.2. *Balance Calculations Results*

The results of the balance calculations for the two gasification experiments conducted are showed in Table 3. Approximately 730 kg and 790 kg of raw lignite feed were consumed during the gasification tests with the Velenje and Oltenia samples, respectively. The remaining part the coal feed was left in the reactor as a gasification char, or unreacted, dried coal. This was confirmed by the post-gasification examination of the UCG cavity.

Table 3. Balance calculations for the Velenje and Oltenia atmospheric pressure experiments.

Parameter	Velenje	Oltenia
Total coal consumption (kg)	730	790
Average coal consumption rate (kg/h)	6.1	8.2
Average gas production rate (Nm^3/h)	5.7	6.1
Average reactor power (kW)	10.3	8.1
Gross energy efficiency (%)	44.6	33.4

The study showed that at similar conditions, the gasification of Oltenia lignite took place with a significantly higher coal consumption rate, i.e., 8.2 kg/h compared to 6.1 kg/h for Velenje lignite. These differences can be explained by the different reactivities of the samples used. According to the energy balance calculations, the Velenje process was characterized by a much higher energy efficiency, i.e., 44.6% (calculated as a ratio of the energy in coal-to-the energy output in gas), compared to the gasification of Oltenia lignite (33.4%). The lower energy efficiency obtained for the Oltenia experiment was due to the higher moisture content of the raw lignite sample. The over-stoichiometric moisture content in the lignite sample eliminates the need to add additional reagent water, at least in the early stage of the gasification. However, excess water may result in a poor gasification efficiency, due to significant heat losses for the evaporation of coal moisture. The obtained results suggest that the appropriate length of gasification channels may be a key issue in improving the efficiency of the process in lignite seams. Recovering the heat of the gas at the surface is an option that can be considered to improve the energy efficiency of the process. This issue may be particularly important in the case of Oltenia lignite, the gasification of which resulted in a significantly lower energy efficiency.

The maximum energy efficiency of approximately 45%, obtained during the gasification of the Velenje lignite sample, was still much less than the values that are typical for UCG of hard colas, i.e., 60–70%.

3.3. Temperature Distribution

The distributions of temperatures over the course of the experiment in the gasification channel, and at two heights from the bottom of the seam are presented in Figures 9 and 10.

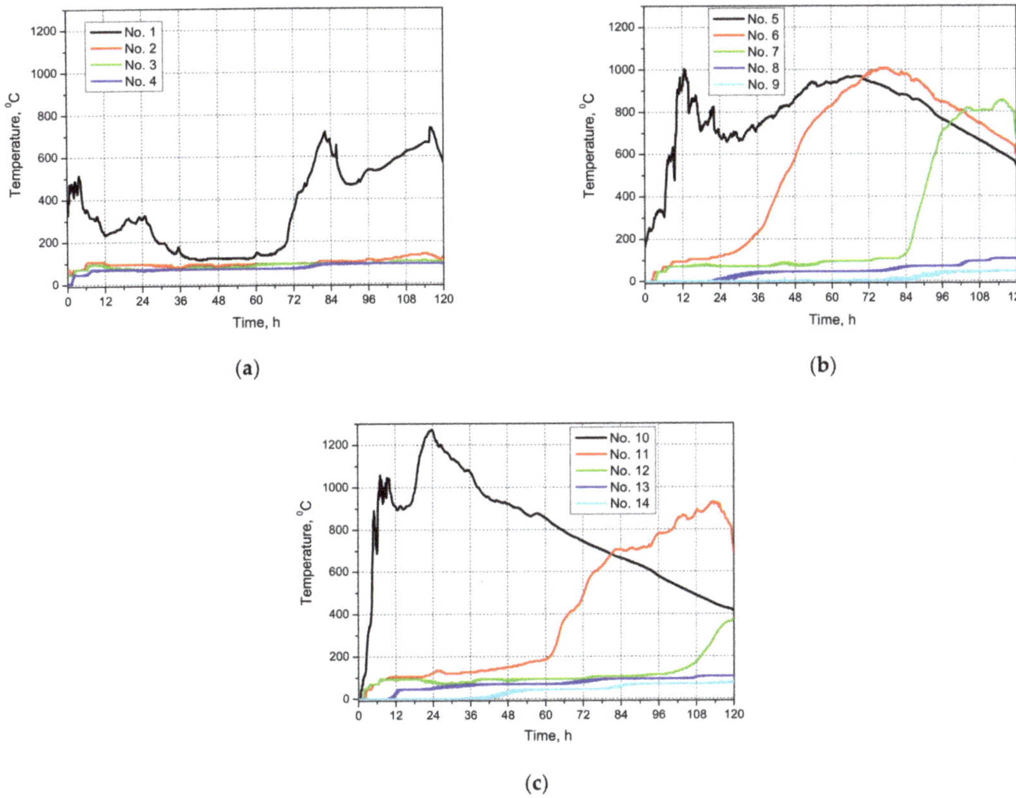

Figure 9. Temperature distribution during the Velenje gasification experiment: (**a**) gasification channel, (**b**) 0.3 m above bottom, (**c**) 0.6 m above bottom.

The maximum temperature during the Velenje test was approximately 1300 °C, and it was recorded at 0.3 m above the gasification channel, after approx. 24 h of the gasification run. The analysis of the temperatures indicates that during the whole course of the experiment, thermal conditions promoting the water gas reaction, as well as the Boudouard reaction (>750 °C), were achieved. Another observation that may be drawn from Figure 9 is that the temperatures in the bottom strata (gasification channel) were about 200 to 400 °C lower than the temperatures in the upper levels of the seam. This observation confirms that the gasification residues (ash and slag) may effectively insulate against heat conduction to the bottom strata, during the UCG process. For the Oltenia experiment, the maximum gasification temperature was about 1380 °C, and it was detected at 0.3 m above the gasification channel, at the early stage of the process (Figure 10). Similarly to the Velenje UCG experiment, during the whole course of the experiment, thermodynamic conditions promoting the water gas and Boudouard reactions were achieved.

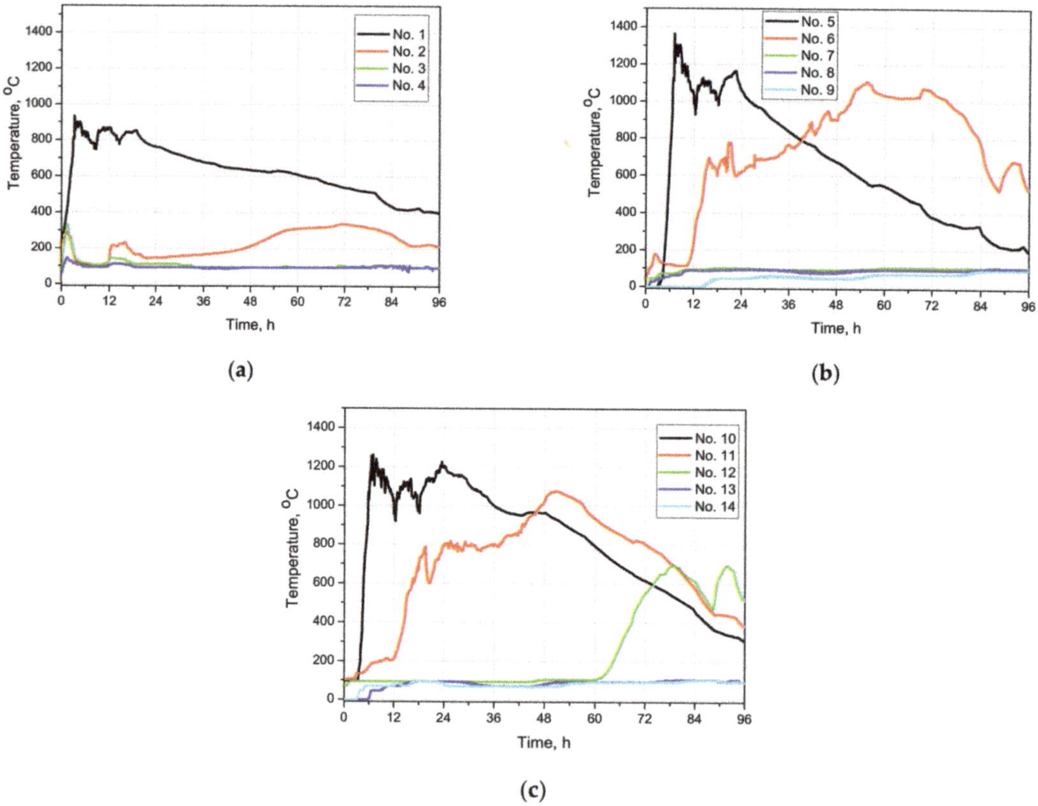

Figure 10. Temperature distribution during the Oltenia gasification experiment: (**a**) gasification channel, (**b**) 0.3 m above bottom, (**c**) 0.6 m above bottom.

4. Conclusions

The gasification tests that were conducted demonstrated that the physicochemical properties of the lignite used considerably affect the in situ gasification process. For similar process conditions (coal seam dimensions, oxygen supply rates), the gasification of Velenje lignite resulted in considerably better gas quality and process efficiency. The overall energy efficiencies, expressed as a ratio of energy in the obtained gas-to-the energy in the coal feed consumed, in the Velenje experiment, were significantly higher compared to the Oltenia test. The main reason behind this is the over-stoichiometric moisture content in the Oltenia lignite sample. The results of the study indicate that underground gasification may be a feasible option for the extraction of lignite deposits, especially in the case of Velenje lignite, which was characterized by a relatively higher calorific value, and lower moisture and ash content.

Funding: This research was funded by the EU Research Fund for Coal Steel, grant number RFCR-CT-2014-00003 and 00774—MEGAPlus—RFCS-201. The work was also partly supported by the Polish Ministry of Science and Higher Education.

Institutional Review Board Statement: Not applicable.

Informed Consent Statement: Not applicable.

Data Availability Statement: Not applicable.

Acknowledgments: The author is thankful to project coordinator—Institute for Studies and Power Engineering (ISPE), Romania and project partners: Complexul Energetic Oltenia and Premogovnik Velenje DD for lignite samples preparation.

Conflicts of Interest: The author declares no conflict of interest.

References

1. International Energy Agency. *World Energy Outlook 2019*; IEA: Paris, France, 2019; Available online: https://www.iea.org/reports/world-energy-outlook-2019 (accessed on 3 August 2021).
2. Perkins, G. Underground Coal Gasification. Part I: Field demonstrations and process performance. *Progr. Energy Comb. Sci.* **2018**, *67*, 158–187. [CrossRef]
3. Couch, G. *Underground coal Gasification*; CCC/151; IEA Clean Coal Centre: London, UK, 2009; p. 129.
4. Bhutto, A.W.; Bazmi, A.A.; Zahedi, G. Underground coal gasification: From fundamentals to applications. *Prog. Energy Combust. Sci.* **2013**, *39*, 189–214. [CrossRef]
5. Zagorščak, R.; An, N.; Palange, R.; Green, M.; Krishnan, M.; Thomas, H.R. Underground coal gasification—A numerical approach to study the formation of syngas and its reactive transport in the surrounding strata. *Fuel* **2019**, *253*, 349–360. [CrossRef]
6. Klebingat, S.; Kempka, T.; Schulten, M.; Azzam, R.; Fernández-Steeger, M.T. Optimization of synthesis gas heating values and tar by-product yield in underground coal gasification. *Fuel* **2018**, *229*, 248–261. [CrossRef]
7. Klebingat, S.; Kempka, T.; Schulten, M.; Azzam, R.; Fernandez-Steeger, T.M. Innovative thermodynamic underground coal gasification model for coupled synthesis gas quality and tar production analyses. *Fuel* **2016**, *183*, 680–686. [CrossRef]
8. Perkins, G. Underground coal gasification—Part II: Fundamental phenomena and modeling. *Prog. Energy Combust. Sci.* **2018**, *67*, 234–274. [CrossRef]
9. Perkins, G.; Sahajwalla, V. A Mathematical Model for the Chemical Reaction of a Semi-infinite Block of Coal in Underground Coal Gasification. *Energy Fuels* **2005**, *19*, 1679–1692. [CrossRef]
10. Laciak, M.; Kostúr, K.; Durdán, M.; Kačur, J.; Flegner, P. The analysis of the underground coal gasification in experimental equipment. *Energy* **2016**, *114*, 332–343. [CrossRef]
11. Burton, E.; Friedmann, J.; Upadhye, R. *Best Practices in Underground Coal Gasification*; Lawrence Livermore National Laboratory: Livermore, CA, USA, 2005.
12. BP Statistical Review of World Energy 2019. Available online: https://www.bp.com/content/dam/bp/business-sites/en/global/corporate/pdfs/energy-economics/statistical-review/bp-stats-review-2019-coal.pdf (accessed on 25 July 2021).
13. Bielowicz, B.; Kasiński, J.R. The possibility of underground gasification of lignite from Polish deposits. *Int. J. Coal Geol.* **2014**, *131*, 304–318. [CrossRef]
14. Kapusta, K.; Wiatowski, M.; Stańczyk, K. An experimental ex-situ study of the suitability of a high moisture ortho-lignite for underground coal gasification (UCG) process. *Fuel* **2016**, *179*, 150–155. [CrossRef]
15. Kapusta, K.; Stańczyk, K.; Wiatowski, M. Comparison of the Contaminants in the Wastewater Produced in the Ex Situ Underground Ortho- and Meta-Lignite Gasification. *Water Air Soil Pollut.* **2019**, *230*, 200. [CrossRef]
16. Strugała–Wilczek, A.; Basa, W.; Kapusta, K.; Soukup, K. In situ sorption phenomena can mitigate potential negative environmental effects of underground coal gasification (UCG)—An experimental study of phenol removal on UCG-derived residues in the aspect of contaminant retardation. *Ecotoxicol. Environ. Saf.* **2021**, *208*, 111710. [CrossRef] [PubMed]
17. Kapusta, K.; Stańczyk, K. Chemical and toxicological evaluation of underground coal gasification (UCG) effluents. The coal rank effect. *Ecotoxicol. Environ. Saf.* **2015**, *112*, 105–113. [CrossRef] [PubMed]
18. Stańczyk, K.; Howaniec, N.; Smoliński, A.; Świądrowski, J.; Kapusta, K.; Wiatowski, M.; Grabowski, J.; Rogut, J. Gasification of lignite and hard coal with air and oxygen enriched air in a pilot scale ex situ reactor for underground gasification. *Fuel* **2011**, *90*, 1953–1962. [CrossRef]
19. Stańczyk, K.; Smoliński, A.; Kapusta, K.; Wiatowski, M.; Świądrowski, J.; Kotyrba, A.; Rogut, J. Dynamic experimental simulation of hydrogen oriented underground gasification of lignite. *Fuel* **2010**, *89*, 3307–3314. [CrossRef]
20. Wiatowski, M.; Kapusta, K.; Stańczyk, K. Efficiency assessment of underground gasification of ortho- and meta-lignite: High-pressure ex situ experimental simulations. *Fuel* **2019**, *236*, 221–227. [CrossRef]
21. Gür, M.; Eskin, N.; Okutan, H.; Arısoy, A.; Böke, E.; Altıntaş, Ü.; Büyükşirin, A.Y.O.; Canbaz, E.D.; Yıldırım, O. Experimental results of underground coal gasification of Turkish lignite in an ex-situ reactor. *Fuel* **2017**, *203*, 997–1006. [CrossRef]

energies

MDPI

Article

Experimental Studies of the Effect of Design and Technological Solutions on the Intensification of an Underground Coal Gasification Process

Oleg Bazaluk [1], Vasyl Lozynskyi [2,*], Volodymyr Falshtynskyi [2], Pavlo Saik [2], Roman Dychkovskyi [2] and Edgar Cabana [3]

1 Belt and Road Initiative Institute for Chinese-European Studies (BRIICES), Guangdong University of Petrochemical Technology, Maoming 525000, China; bazaluk@ukr.net
2 Department of Mining Engineering and Education, Dnipro University of Technology, 49005 Dnipro, Ukraine; falshtynskyi.nmu@gmail.com (V.F.); saik.nmu@gmail.com (P.S.); Dychkovskyi.r.o@nmu.one (R.D.)
3 Institute of the Center of Renewable Energy and Energy Efficiency, Universidad Nacional de San Agustin de Arequipa, Arequipa 04000, Peru; ecaceresca@unsa.edu.pe
* Correspondence: lvg.nmu@gmail.com

Citation: Bazaluk, O.; Lozynskyi, V.; Falshtynskyi, V.; Saik, P.; Dychkovskyi, R.; Cabana, E. Experimental Studies of the Effect of Design and Technological Solutions on the Intensification of an Underground Coal Gasification Process. *Energies* **2021**, *14*, 4369. https://doi.org/10.3390/en14144369

Academic Editors: Ján Kačur, Milan Durdán and Marek Laciak

Received: 21 June 2021
Accepted: 15 July 2021
Published: 20 July 2021

Abstract: This paper represents the results of experimental studies of physical modeling of the underground coal gasification process in terms of implementation of design and technological solutions aimed at intensification of a gasification process of thin coal seams. A series of experimental studies were performed in terms of a stand unit with the provided criteria of similarity to field conditions as well as kinetics of thermochemical processes occurring within a gas generator. Hard coal (high volatile bituminous coal) was selected as the raw material to be gasified, as that coal grade prevails in Ukrainian energy balance since it is represented by rather great reserves. Five blow types were tested during the research (air, air–steam, oxygen–steam, oxygen–enriched, and carbon dioxide and oxygen). As a result, the effect of tightness of a gas generator on the quantitative and qualitative parameters of coal gasification while varying the blow by reagents and changing the pressure in a reaction channel has been identified. Special attention was paid to the design solutions involving blow supply immediately into the combustion face of a gas generator. The experimental results demonstrate maximum efficiency of the applied gas generator design involving flexible pipelines and activator in the reaction channel and a blow direction onto the reaction channel face combined with blow stream reversing which will make it possible to improve caloricity of the generator gas up to 18% (i.e., from 8.4 to 12.8 MJ/m^3 depending upon a blow type). Consideration of the obtained results of physical modelling can be used with sufficient accuracy to establish modern enterprises based on the underground coal seam gasification; this will help develop more efficiently the substandard coal reserves to generate heat energy as well as power-producing and chemical raw material. The research conclusions can provide technical reference for developing a new generation of underground coal gasification technology.

Keywords: coal gasification; rocks; coal seam; material balance; heat balance; tightness; gas

1. Introduction

Currently, different studies are being carried out analyzing deep coal processing aimed both at manufacturing of energy products for electric energy generation and obtaining valuable chemical products [1–3]. Use of coal to generate syngas, methanol, liquid fuel, and other deficit products is the tendency of special topicality [4–6]. That makes it possible to consider coal as a reliable alternative source of obtaining carbohydrate raw material, especially in terms of exhaustion of oil and gas reserves due to growing volumes of their consumption and low rates of additional exploration of oil-and-gas fields [7–11].

Currently, production of energetically valuable liquid fuel from coal is the industrially developed process; in this context, reactions of incomplete coal oxidation are quite often to have in surface gas generators [12,13]. A main disadvantage of surface gas generators is their high cost and considerable expenditures for coal extraction and transportation to the place of processing [14]. Underground coal gasification (UCG) is a prospective tendency of deep coal processing; that provides the reactions of incomplete coal oxidation in terms of underground conditions immediately within the place of coal seam occurrence—in an underground gas generator—with production of gas (after its corresponding surface-based processing), which is close to natural gas in its consumer properties [15,16]. Moreover, the gas of underground coal gasification may be the raw material for getting syngas, methanol, ammonia, carbamide, and other valuable chemical products [17–22]. At the same time, it is critically important to provide environmental safety during mining [23–27].

Underground gasification of solid fuel is an important tendency in the development of natural fuel deposits; it means underground fuel transformation into a combustible gas for its further energetic and technological use after its outlet to the earth's surface [28]. Main feature of underground coal gasification is that such mining method helps develop both off-balance and non-commercial coal reserves [29–32]. The non-traditional mining method of coal deposits opens new prospects in the development of coal seams with complicated mining and geological modes of occurrence. It combines extraction, dressing, and complex processing [33–35] with detailed study of rock mass behavior [36–41].

The essence of the technology of underground coal gasification is in drilling of wells from the earth's surface towards a coal seam by means of directed drilling, their linkage within a seam by one of the known techniques with further coal seam ignition, creation of the conditions for coal transformation right underground into a combustion gas, and removal of the produced gas through the wells onto the earth's surface [42–44].

Advantages of the mentioned technology are as follows: coal is not hoisted to the surface, there are no large rock volumes to be placed somewhere, the terrain integrity is not disturbed, there is no subsidence due to formation of underground voids, and there is no need in using additional chemical reagents that have negative impact on the environment [45–47]. Thus, all the technological operations of underground coal gasification are performed from the earth's surface without miners' underground operations. The technology belongs to so-called clean coal technologies, it is environmentally safe, and it can become a great alternative to extraction [48–52].

A process of underground coal gasification (UCG) consists of the primary chemical reactions within an oxidizing zone, a reaction channel of a gas generator [53]. A combustion (oxidizing) process in that zone is accompanied by the formation of carbon monoxides and dioxides. A reducing zone of a gas generator shows the processes accompanied by the secondary reducing reactions with the formation of gasification products [54,55]. A degree of coal transformation into gasification products depends on certain key conditions and parameters: ultimate composition, texture and structure of a coal seam and roof and floor rocks, hydrogeological conditions, pressure, temperature, composition of a blow mixture, directionality and duration of blow contact in space and time with a combustion face of an underground gas generator [56,57].

Currently, there is sufficient number of studies analyzing the features of design and technological solutions in terms of expediency of underground gas generator tightness [58–61], methods of supply and directionality of blow flows with the provided adaptivity and intensification of the processes of coal seam gasification [62–65] including controlled retracting injection point (CRIP) system [66,67], influence of mining and geological conditions of coal seam occurrence on the underground gasification process [68–72], and influences of underground coal gasification on the environment [46,58,73–76].

Application of different technical means, methods, and techniques as well as physical fields makes it possible to intensify the UCG process by effecting two objects: immediate coal mass and blow flows injected into the combustion face of an underground gas generator [77–82].

Tightness of the gas generator design is provided by injection-stowing operations within the deformed layers of the rocks enclosing the seam [83]. Stowing operations are performed with the consideration of changes in geomechanical parameters, parameters of a temperature field of the "rock mass—gas generator" system, changes in the advance rate of a combustion face of a gas generator, and increase in the gasified area in time and space [84–86]. While using injection stowing in terms of gasification of thin coal seams, ground subsidence is not more than 11–18% of the coal seam thickness [87].

Introduction of injection stowing of the deformed natural roof rock and the gasified space within the UCG stations in the underground gas generator design will ensure the mobility of a coal seam gasification process owing to the expansion of the area for the technology implementation. Artificial leak-tightening of a gas generator helps increase the criteria of controllability, compactness, environmental credentials, and safety of the process, which will allow performing the underground coal gasification process in the adapted modes of pressure and temperatures. That stipulates the growth of quantitative and qualitative indices of the UCG process since a set of factors, among which increased pressure in a reaction channel and features of the underground gas generator design are the key ones, results in the considerable losses of blow and generator gas during the coal seam gasification [88,89].

The research objective is to carry out the experimental modeling and study the effect of design and technological solutions, which influence immediately the degree of efficiency of the underground coal gasification process.

To reach the objective the authors, based on the experimental studies, strive to identify the effect of gas generator tightness on the quantitative and qualitative parameters of coal gasification (loss of blow and gas; changes in heat conductivity, temperature, and combustion heat as well as in the output of combustible generator gases and chemical products) in terms of blow variation by reagents and changing pressure in a reaction channel as well as to determine the effect of design and technical solutions for blow supply immediately onto the combustion face of a gas generator (loss of blow and gas, pressure, coal, advance rate and time that the main processes are achieving) depending on the type of gas generator design.

2. Materials and Methods

2.1. Determining the Sufficient Tightness of a Gas Generator

While substantiating the sufficient tightness of an underground gas generator, the following was considered:

(1) Permeability of the coal-overlaying thickness taking into account natural and artificial fractures of the coal-overlaying thickness without injection stowing of the deformed rocks is determined according to the formula (1):

$$K_{c.o.t} = \sum_{i=1}^{n} \frac{k_p^i \cdot h_{r.l}^i \cdot P_{r.r}}{h_{r.c} \cdot P_{r.c}}, \quad (1)$$

where $K_{c.o.t}$ is coefficient of permeability taking into account natural and artificial fractures, of the coal-overlaying thickness before stowing operations: k_p^i is coefficient of permeability taking into account natural and artificial fractures of the rock layers of the roof before stowing operations; $h_{r.l}^i$ is thickness of rock layers of the roof, m; $P_{r.r}$ is pressure within the roof rocks, MPa; $h_{r.c}$ is thickness of reaction channel, m; and $P_{r.c}$ is pressure within the reaction channel of a gas generator, MPa.

(2) Permeability of the coal-overlaying thickness taking into account natural and artificial fractures of the coal-overlaying thickness after injection stowing of the deformed rocks is determined according to the formula (2):

$$K_{c.o.t.s} = \sum_{i=1}^{n} \frac{k_{p.s}^i \cdot h_{r.l}^i \cdot P_{s.r.r}}{h_{r.c} \cdot P_{r.c}}, \quad (2)$$

where $K_{c.o.t.s}$ is coefficient of permeability taking into account natural and artificial fractures of the coal-overlaying thickness after stowing operations; $k_{p.s}^i$ is coefficient of permeability taking into account natural and artificial fractures of the rock layers of the roof after stowing operations; $h^i_{r.l}$ is thickness of rock layers of the roof, m; $P_{s.r.r}$ is pressure within the stowed roof rocks, MPa; $h_{r.c}$ is thickness of reaction channel, m; and $P_{r.c}$ is pressure within the reaction channel of a gas generator, MPa.

Coefficient of sufficient tightness of an underground gas generator $K_{t.u.g/g}$ with the consideration of the roof rock permeability during injection stowing and temperature effect on the mass under stowing is to be identified according to the formula (3):

$$K_{t.u.g/g} = \frac{K_{c.o.t.s} \cdot \beta_{e.m}}{K_{c.o.t}}, \quad (3)$$

where $K_{c.o.t}$ is coefficient of permeability taking into account natural and artificial fractures of the coal-overlaying thickness before stowing operations; $K_{c.o.t.s}$ is coefficient of permeability taking into account natural and artificial fractures of the coal-overlaying thickness after stowing operations; and $\beta_{e.m}$ is coefficient of temperature expansion of the mass under stowing.

2.2. Determining the Effect of Heat Exchange

A degree of the effect of convective heat exchange within the rocks enclosing a gas generator is determined by density, structure, and dimensions of the voids. General amount of heat (Q) transferred within a certain time period through a rock layer h_r is determined according to the formula (4):

$$Q = \left(\frac{\lambda_{eq}}{h_r} - \alpha_i\right) \cdot (T_1 - T_2), \quad (4)$$

where: α_i is coefficient of heat conductivity of the rocks; h_r is thickness of the rock layer; T_1 is initial temperature; T_2 is final temperature; and λ_{eq} is equivalent coefficient of heat conductivity of the rocks is determined according to the formula (5):

$$\lambda_{eq} = \lambda_r \cdot (1 - P)^3, \quad (5)$$

where: λ_r is coefficient of heat conductivity of the rocks, if there are no artificial fractures in the rocks (intact rock mass); P is value of natural and artificial fractures in the rocks.

Heat conductivity of the rocks is in cubic dependence on the rock porosity and fractures. The stowing of natural and technogenic voids within the deformed layers of the roof results in changes in heat conductivity from the convective to conductive one.

Density of the rock mass enclosing a gas generator decreases due to the effect of rock pressure, which experiences its changes in time and space along with the combustion face advance and stipulates the prevailing convective heat exchange within the roof rocks.

2.3. Experimental Studies

The studies were carried out to test the technological schemes of underground coal gasification, if the gas generator design is changed, i.e.:

- Gas generator design with stowing of the deformed thickness of the roof rocks and the gasified space;
- Gas generator design without stowing.

Apart from the tightness of a gas generator, technical solutions of the blow reagent supply into the combustion face were the subject of experimental identification of the optimal method of the gasification process performance. Following solutions of the blow reagent supply were tested:

- Without a flexible pipeline with the blow direction onto the reaction channel face;
- With flexible pipelines with the blow direction through perforated nozzles onto the reaction channel face;

- With flexible pipelines and activator in the reaction channel, with the blow direction onto the reaction channel face.

2.3.1. Experimental Stand Unit

The experimental stand unit is designed and patented at Dnipro University of Technology, and manufactured by Naftomash RMA under financial support of the Ministry of Education and Science of Ukraine.

A gas generator model consists of four systems (Figure 1):

- An experimental stand;
- A system of supply of separated and mixed blow mixture (blow reagents, see Section 2.3.3);
- A gas outlet system;
- A system of control and measuring equipment (temperature control and control of input and output gas mixtures).

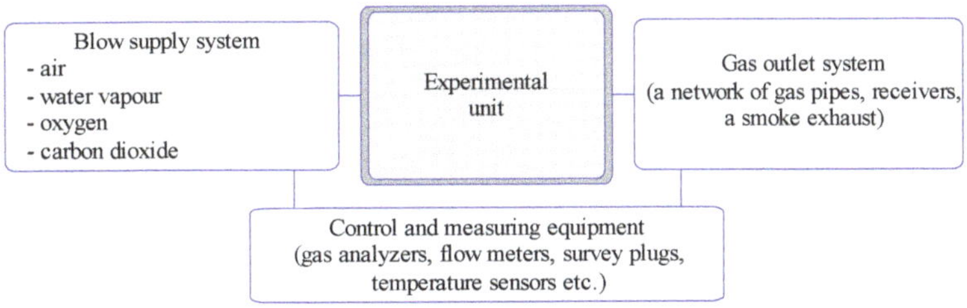

Figure 1. Technological scheme of a stand unit of underground coal gasification.

An experimental stand welded from sheet steel is a central link of the facility. There are holes for blow supply and generator gas outlet as well as ignition and control holes on the stand front [90,91]. A system of thermal sensors with signal converters equipped with the interface was used to identify the temperature field parameters. The reference shear detectors were used to control the coal mass state.

The coal rock mass was modelled in terms of the experimental unit according to the similarity criteria (see Section 2.3.5).

2.3.2. Ultimate and Technical Composition of Coal

Hard coal (high volatile bituminous coal) was selected as the raw material to be gasified. That coal grade prevails in Ukrainian energy balance as it is represented by rather great reserves. Table 1 shows proximate and ultimate analysis of the coal.

Table 1. Proximate and ultimate analysis of the coal.

Proximate Analysis				Ultimate Analysis					Combustion Heat (Q_r), MJ/kg	Coal Density (γ), g/cm^3
W^r, %	W^a, %	A^c, %	S^d, %	V^{daf}, %	C^{daf}, %	H^{daf}, %	O^{daf}, %	Nr, %		
1.7	2.2	38.2	1.3	37.0	80.7	6.3	6.8	4.9	24.6	1.45

The values shown in Table 1 are used to calculate the material and heat balance (see Section 2.3.4), being an indispensable part of research as the obtained results of calculations define the required amount of blow to be supplied into a gas generator.

2.3.3. Blow Reagents

The stand-based experimental studies were accompanied by changes in the parameters of blow reagents. Five basic previously tested blow types were used, i.e.:

I. Air blow (O_2—21%, N_2—79%);

II. Air–steam blow (O_2—21%, N_2—79%, H_2O_{steam});
III. Oxygen–steam blow (O_2—35%, N_2—65%, H_2O_{steam});
IV. Oxygen–enriched blow (O_2—35%, N_2—65%);
V. Carbon dioxide and oxygen (O_2—21%, CO_2—10%, N_2—69%).

A series of experimental studies was alternated by means of blow change. Each following change in blow mode was followed by the transfer to the air blow. A transition mode in terms of blow lasted one hour.

2.3.4. Material and Heat Balance

The material and heat balance are determined by physical rates of chemical reactions, technological efficiency of the process, and modes of blow mixture supply into the gasification zone [92]. To calculate the material and heat balance, the MT-Balance software was applied. The software product was developed at the Dnipro University of Technology [93,94].

During the calculation, the software uses not only the specified parameters of the ultimate and technical composition of the coal but also many physical values-constants as well as the values characterizing the initial state of the gasification process. The software algorithm makes it possible to obtain the following:

- Material balance of the oxidizing zone;
- Material balance of the reducing zone;
- Volumetric parameters of gas mixtures of a gas generator;
- Chemical and physical efficiency of the gasification process;
- Energy balance of the gasification process;
- Total energy of the oxidizing and reducing zones.

Identification of the material and heat balance with the help of underground gasification is a valid and convenient mechanism for obtaining quantitative and qualitative parameters of the blow mixture composition and gases outgoing from a gas generator. That allows simplifying considerably the data processing and helps obtain rapidly the final results with high degree of conformity.

2.3.5. Similarity Criteria

Modeling of the operating parameters of the process of underground coal gasification according to the criteria of similarity to field conditions are the important elements of the experimental data transfer into the field conditions as the performance of experiments in terms of ground stand units are aimed at simulation of the UCG processes taking into account geological and technological parameters [95,96].

The research was carried out in terms of the experimental stand unit to model a process of underground gasification taking into consideration the similarity criteria and the specified scale coefficients. Following expression is taken as the basis in terms of non-stationary seam gasification [97]:

$$\overline{T} = \frac{T - T_0}{T_{max} - T_0} = f(H_0), \tag{6}$$

where T is current temperature, °C; T_{max} is maximum temperature, °C; T_0 is initial temperature, °C; H_0 is criterion of homochronicity (of time).

$$H_0 = \frac{v \cdot t}{x} = 1, \tag{7}$$

where v is gasifying rate, m/day; t is gasifying time, days; x is distance, m.

Generally, all similarity criteria, taking into consideration time, are called homochronicity criteria (*Ho*) since they are applied to identify a time conversion factor through a multiplier of other physical quantities. Hence, in terms of similarity of two or more systems, *Ho* (homochronicity criterion), *Fr* (Froude number), Eu (Euler's criterion), and *Re* (Reynolds number) have the same values for any similar points. In practice of similarity criteria use, it is expedient to reduce some of them to more convenient format helping determine directly values being a part of the criteria.

Thus, a conclusion concerning full nature-model similarity compliance makes it possible to calculate homochronicity criterion, i.e., constant temporal similarity within the processes. Comparison of modeling of working parameters of underground coal gasification helps obtain the following:

$$Ho = \frac{v \cdot t}{x} = \frac{v' \cdot t'}{x'}, \tag{8}$$

where v is actual displacement velocity of a material point, m/day; t is displacement time of the point, days; x is distance passed by the material point during t time, m; and v', t', x' are velocity, time, and path of a similar material point on the model, respectively.

To obtain the valid results of the modelling in terms of the experimental stand unit that would help get the data for field conditions, a group of similarity invariants, characterizing the gasification process, were considered [98]:

- Kinetics of chemical reactions;
- Gas dynamics and mass exchange of the oxidizing and reducing zones;
- Convective and conductive heat exchange.

Adherence to the abovementioned similarity invariants, representing the gasification kinetics, was obtained by the fact that the coal model was of the same grade and composition as the "field" coal.

Convectional temperature exchange takes place right within a contact of reactional zone expanding up to 2–4 thicknesses of the degassed seam. Then, enthalpy takes place at the expense of conductive rock heating. Temperature rise and its expansion deep in the rock mass last until origination of thermal stresses varying rock behavior while falling [99]. Based upon previous research carried out under the conditions of Western Donbas mines, it can be concluded that convectional enthalpy transfer was observed at greater distances than 6 m [100]. This situation is justified by the stratification cavities formation in the roof of an underground gasifier and the presence of pores and fractures in it. The degree of influence of heat exchange convection is determined by the presence of cavities in the rocks and their location, which in turn determines the thermal stress of the adjacent roof rocks.

The pressure in the underground gas generator model and the ultimate composition of a generator gas in the similarity scale meet the field conditions. According to the calculated data, air consumption during the air blow was 2.32 m^3/kg of coal; in case of blow enriched with oxygen, it was 1.89 m^3/kg of coal. The reaction channel length within the experimental stand is 1.5 m; thickness of the coal seam is 0.5 m. Taking into account the similarity criteria, a 1.5 m value of the reaction channel length corresponds to the reaction channel length of the underground gas generator being 30 m.

3. Results and Discussion

As a result of the stand-based experimental studies, the influence of tightness characteristics of a gas generator on the quantitative and qualitative parameters of coal seam gasification have been identified. Figure 2 demonstrates a graph of changes in losses of the blow supplied into a gas generator and the obtained generator gas in terms of changing pressure growth from 0 to 0.35 MPa.

The Graph demonstrates that stowing nonavailability starts impacting blow and generator gas losses in terms of minimum pressure change achieving maximum 32.5% values at 0.35 MPa pressure. In turn, at the same pressure value, stowing is only 15%. Hence, 17.5% (i.e., more than double) difference in blow losses has been identified depending upon various modes of preparation of gas generators.

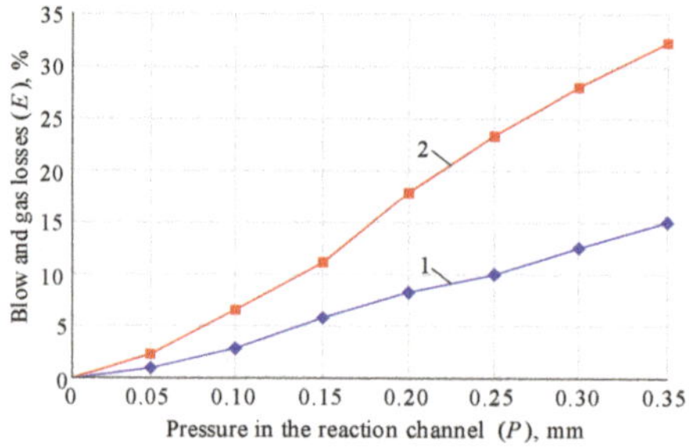

Figure 2. Dependence of blow and gas losses (*P*, %) in terms of pressure changes in an underground gas generator: 1—gas generator design with stowing of the deformed thickness of the roof rocks and gasified space; 2—gas generator design without stowing.

A gas generator design with stowing of the deformed thickness of the roof rocks and gasified space is characterized by $K_{t.u.g/g}$ = 0.032 coefficient of leak-tightness. At the same time, the tightness coefficient of a gas generator without stowing is $K_{t.u.g/g}$ = 0.214.

In terms of sufficient gas generator tightness ($K_{t.u.g/g}$ = 0.032), it is possible to ensure effective contact of blow and a combustion face of the reaction channel. In addition, heat capacity of the rocks enclosing a gas generator increases; heat efficiency of the reaction zones of a gas generator grows stipulating stability of material and heat balance of the coal seam gasification process.

Figure 3 shows the dependences of heat conductivity of the rocks around a gas generator on the temperatures emitting during gasification and tightness of a gas generator.

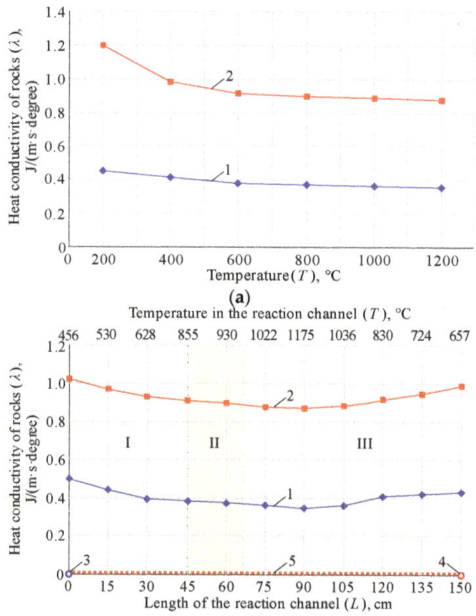

Figure 3. Dependences of heat conductivity of the rocks (λ) around a gas generator on the temperatures emitting during gasification and tightness of a gas generator: (**a**)—general conductivity of the roof rocks of a gas generator; (**b**)—heat conductivity of the roof rocks along the combustion face of a gas generator; 1—generator design with stowing of the deformed thickness of the roof rocks and gasified space; 2—generator design without stowing; 3—air-supply well; 4—gas-outlet well; 5—line of the combustion face of a gas generator; I—oxidizing zone; II—transition zone; III—reducing zone.

Increase in temperatures in the oxidizing zone of a gas generator and heat capacity of rock layers are observed throughout the reaction channel length, which is stipulated by heat losses in the rocks around the reaction channel. That happens at the expense of decreasing heat and temperature conductivity of the roof rocks. Heat-generating capacity of the oxidizing zone at the expense of artificial heat insulation of the rock mass provides efficient operation of the reducing zone, where endothermal reactions with heat absorption occur. Technogenic heat of the rock mass around a gas generator is rather considerable; correspondingly, in terms of its removal, it can be used in cogeneration plants.

Coal is gasified in terms of blow injection from the side of coal mass and gas removal from the gasified space. Along with the expanding degassed void, consumer features (calorific capacity) of gas are deteriorating as the gasification front is displacing gradually from the initial location to the gasification boundaries, leaving behind the gasified space, which is filled gradually with slugs, residual coal, and deposited rocks of the upper thickness. The fly ash is driven out of coal from an underground gasifier together with the generator gases. It is generally captured by particle filtration equipment before the flue gases reach the chimneys. Non-reacted coal ash is left in the mined-out space. The volume of non-reacted coal ash makes up 38–45% of the volume of the coal seam.

Figure 4 shows the results of analyzing the temperatures of gasification products at their output from the gas-outlet well and combustion heat of a generator gas of a thin carboniferous seam.

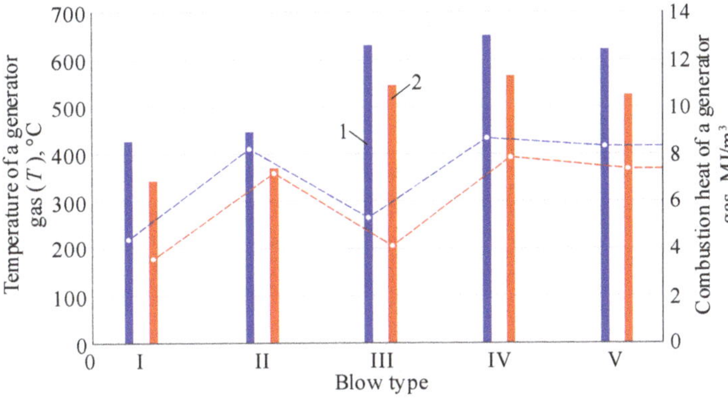

Figure 4. Dependence of changes in the temperature of coal gasification products (generator gas) at their output from a gas-outlet well and combustion heat of a generator gas upon the blow type and tightness of a gas generator: 1—generator design with stowing of the deformed thickness of the roof rocks and gasified space; 2—generator design without stowing; I—air blow (O_2—21%, N_2—79%); II—air-steam blow (O_2—21%, N_2—79%, H_2O_{steam}); III—oxygen-steam blow (O_2—35%, N_2—65%, H_2O_{steam}); IV—oxygen-enriched (O_2—35%, N_2—65%); V—carbon dioxide and oxygen (O_2—21%, CO_2—10%, N_2—69%).

The obtained results of stand-based experimental studies of design features of artificial tightness of a gas generator have made it possible to evaluate the quantitative and qualitative parameters of the output of coal seam gasification products taking into account mining-geological conditions, design features of a gas generator, and technological parameters of the gasification process (Figure 5).

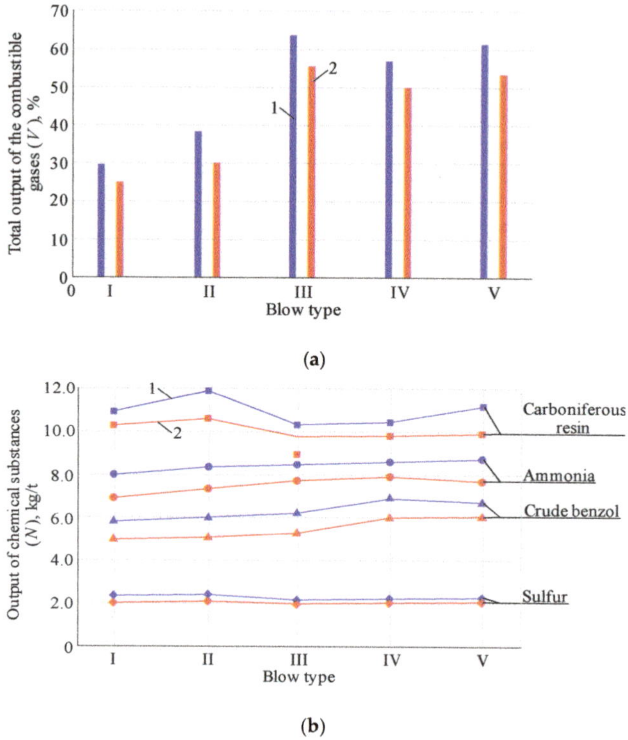

Figure 5. Dependences of changes in the output of coal gasification products (generator gas) upon the blow type and tightness of a gas generator: (**a**)—output of combustible gases; (**b**)—output of chemical substances from a condensate; 1—generator design with stowing of the deformed thickness of the roof rocks and gasified space; 2—generator design without stowing; I—air blow (O_2—21%, N_2—79%); II—air-steam blow (O_2—21%, N_2—79%, H_2O_{steam}); III—oxygen-steam blow (O_2—35%, N_2—65%, H_2O_{steam}); IV—oxygen-enriched (O_2—35%, N_2—65%); V—carbon dioxide and oxygen (O_2—21%, CO_2—10%, N_2—69%).

Decreasing natural and artificial fractures of the layered roof rock thickness and the gasified space at the expense of increased tightness of a gas generator helps reduce migration of high-temperature UCG products into the roof and floor rocks. In turn, that will allow further expansion of the UCG application area in terms of thin coal seams occurring at shallow depths. The calculations have shown that preparation of underground gas generators with injection stowing of the roof rocks will make it possible to increase the gasified coal area by 1.6–1.8 times, the degree of coal seam gasification will grow by 7–12%, and the heat-generation gas capacity will go up to 18%.

At the same time, great attention should be paid to continuous supply of air mixture onto the combustion face of the gas generator's reaction channel as that creates stable pressure and activates heat-generation within the oxidizing zone of the reaction channel. That results in the balanced behavior of active zones of a gas generator and balanced kinetics of thermochemical reactions of the gasification process.

Having been developed and tested in terms a stand unit for simulation of the underground coal gasification process, the design and technical solutions aimed at blow supplying immediately onto the combustion face stipulate intensification of a gasification process of thin coal seams taking into account specific mining-geological conditions.

Consequently, three possible gas generator designs were to be analyzed as for the efficiency of a gasification process:

(1) Without a flexible pipeline for blow direction onto the reaction channel face (design A);
(2) With flexible pipelines for blow direction through perforated nozzles onto the reaction channel face (design B);
(3) With flexible pipelines and activator in the reaction channel, with blow direction onto the reaction channel face (design C).
(4) Figure 6 represents the results of a series of experiment studies.

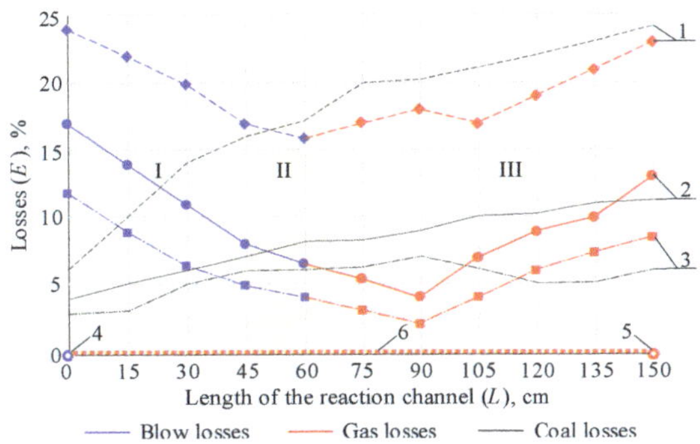

Figure 6. Dependences of the losses of blow, generator gas, and coal upon the gas generator design: 1—design A; 2—design B; 3—design C; 4—air-supply well, 5—gas-outlet well, 6–line of the combustion face of a gas generator; I—oxidizing zone; II—transition zone; III—reducing zone.

If a gas blow flows from the side of mass, gas outlet towards the gasified space is stimulated somehow by large volumes of that space, deformation, and rock caving into the gasification zone as well as the available excessive pressure in it.

Coal is gasified better around the blow well than around the gas-outlet well. Thus, this is the point with higher concentration of coal losses in terms of seam area and thickness. Moreover, it is stipulated not only by the unilateral direction of the blow flow but also by the fact that the advance of reaction zones of gasification along the channel length is accompanied by the deteriorating aerodynamic conditions, and finally there will be a moment when there is no sufficient length for the reaction zones.

Figure 7 summarizes analyses of the effects of different designs of a gas generator (A, B, and C) as for the blow supply into the reaction channel on the uniformity of the combustion face advance.

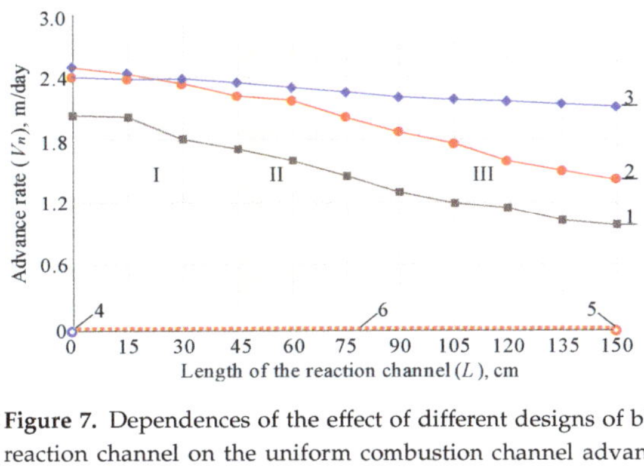

Figure 7. Dependences of the effect of different designs of blow supply into the gas generator's reaction channel on the uniform combustion channel advance depending on the gas generator design: 1—design A; 2—design B; 3—design C; 4—air-supply well, 5—gas-outlet well, 6—line of the combustion face of a gas generator; I—oxidizing zone; II—transition zone; III—reducing zone.

While analyzing Figure 7, it is possible to state the following: if coal seam is gasified involving the gas generator design with flexible pipelines and activator in the reaction channel as well as the blow direction onto the combustion face of a reaction channel (design C), there is no need in blow reverse as we can observe uniform advance of the gas generator's combustion face.

Involvement of the off-balance and abandoned reserves of thin and very thin coal seams into gasification in terms of using the designs and technological schemes of gas generators with the controlled flexible pipelines, perforated nozzles, and activators provides adaptive activation of the oxidizing and reducing processes with the controlled transition zone between them in the reaction channel of a gas generator. Control and controllability of a gas generator throughout the reaction channel length is ensured by the dosed, separate supply of a blow flow onto the combustion face into each active zone of the reaction channel. That results in the reduced losses of blow, generator gas, and solid fuel; it also stipulates uniform advance of active zones of the reaction channel's combustion face during the coal seam gasification.

Implementation of the abovementioned technological solutions in the gas generator designs and technological solutions of coal seam gasification will reduce considerably the time spent for the formation of a reaction channel and beginning of the mine gas generator's operation in active mode of coal gasification (Figure 8).

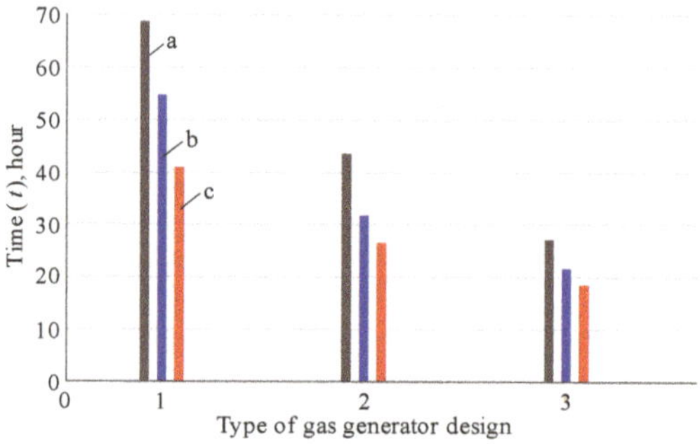

Figure 8. Effect of the gas generator design on the time of achieving the main processes: a—formation of the reaction channel of a gas generator; b—start of the gasification mode of a gas generator; c—reversing; 1—design A; 2—design B; 3—design C.

Consideration of geological structure, ultimate composition of a solid fuel, rocks of the roof and floor as well as the parameters of blow mode and its direction onto the combustion face according to the functional features of active zones of a reaction channel and design solutions of a gas generator stipulates adaptivity of the process of underground coal gasification to a concrete model of thin coal seam development in terms of using flexible pipes with perforated nozzles, or activator.

Control of the process of thin and very thin coal seam gasification with the help of flexible pipelines with blow direction through perforated nozzles onto the reaction channel face (scheme B) does not provide sufficiently effective activation of thermochemical processes within the active zones of a gas generator. In terms of oxidizing zone of the reaction channel, that design solution effects immediately the kinetics of chemical reactions and heat generation. The reducing zone is not affected immediately; thus, the reducing gasification processes proceed under the effect of gas products and heat generated within the oxidizing zone. Figure 8 demonstrates that design C helps reduce almost twofold the time for reaching the main processes, i.e., a gasification process experiences its intensification.

Disadvantages of the design are as follows: impossibility to control the oxidizing zone of a gas generator, nonuniform gasification of the combustion face of the reaction channel that influence negatively the quality of a thermochemical process as well as prolongs the time for preparation and formation of the reaction channel and reverse of blow flows.

Implementation of a perforated activator in the reaction channel of a gas generator to activate the underground coal gasification process ensures immediate effect on active zones of the combustion face; that helps control the active zones maintaining energy balance between the zones and selectivity of a gasification product, and increase quantitative and qualitative parameters of the process. Concerning disadvantages of design C, it can be mentioned that considerable mass of a perforated activator creates resistance during its displacement along with the advance of the gas generator's combustion face. Consequently, critical loads are formed on a flexible pipeline and its connection with the activator causing the need in limitation of the length of the gas generator's extraction pillar.

The obtained results of changes in generator gas output depending upon a blow type are correlated with a high convergence degree with earlier studies of underground coal gasification. Namely, numerous papers mention positive effect by oxygen-enriched blow [101–104] as well as oxygen and steam blow [104,105] on generator gas output indices. Essential effect by carbon dioxide blow has also been mentioned in [106–108]. Moreover, experimental results support confidently earlier studies with the use of similar coal [85]. No doubt, we cannot suggest full compliance of the research results since ultimate and technical analysis of the coal is not identical.

Comparative analysis of the results concerning blow stream reversing with studies by foreign researchers, described scrupulously in [109–111], prove the correctness of the selected tendency to intensify underground gasification process.

As for the comparison of the improved techniques, proposed by this paper, their effect on a gas formation process is incomparable with the similar studies. Nevertheless, we have propose to the studies alternative techniques of blow supply to fire mass of underground gas generator in addition to such popular systems as Controlled Retracting Injection Point (CRIP) [66,112,113], Movable Injection Point [114], Unfixed Pumping Points [115,116], and recently developed gasification agent injection tool for underground coal gasification [116].

The authors believe that further studies will concern analysis of underground gasification effect while degassing thin and very thin coal seams. Namely, determination of the influence of qualitative and quantitative composition of carbonous reagents on the gas formation efficiency within a reaction channel of an underground gas generator is meant. Moreover, a problem to analyze environmental component of a gasification process (among other things, formation of polluting gases and mercaptans and their expansion) remains a topical issue.

4. Conclusions

A series of experimental studies involving the gasification process modeling has helped substantiate the effect of technological and technical innovations in gas generator designs on the parameters of the coal seam gasification process.

Implementation of the designs of underground gas generators with stowing of the deformed roof rocks as well as the ones with the controlled active zones of a gas generator (especially while gasifying the abandoned balance and off-balance reserves of thin and very thin coal seams) provides adaptivity of the coal seam gasification process to specific mining-geological conditions.

A key feature of the innovative solutions of underground coal gasification, stipulating its economic expediency, is creation of favorable conditions for the directed blow supply into the reaction zone of coal mass from the variable points and keeps its contact with coal during the whole planned time of coal gasification.

Preparation of underground gas generators with injection stowing of the roof rocks will help reduce the prime cost of the produced coal by 23.5% owing to the following: area of the gasified coal increases by 1.6–1.8 times; degree of coal seam gasification grows by 7–12%. In this regard, maximum efficiency of underground gasification is achieved if a gas generator design involving flexible pipelines and activator in the reaction channel and blow direction onto the reaction channel face combined with blow stream reversing and oxygen blow use which will make it possible to improve caloricity of the generator gas up to 18% (i.e., from 8.4 to 12.8 MJ/m^3 depending upon a blow type).

Design technological schemes allows controlling efficiently a gas-generation process by directed blow supply and constant forcing of a blow flow against the combustion bed surface by means of combination of gas-blow flows (blow injection and exhaustion of gasification products) and by reversing the blow flows.

Author Contributions: Conceptualization, V.L. and V.F.; methodology, V.L. and V.F.; software, P.S. and V.L.; validation, R.D. and E.C.; formal analysis, O.B. and P.S.; investigation, V.L. and V.F.; resources, P.S.; data curation, E.C.; writing—original draft preparation, V.L. and V.F.; writing—review and editing, R.D.; visualization, V.L. and P.S.; supervision, R.D.; project administration, O.B.; funding acquisition, O.B. All authors have read and agreed to the published version of the manuscript.

Funding: This study was carried out as part of the project "Belt and Road Initiative Institute for Chinese-European studies (BRIICES)" and was funded by the Guangdong University of Petrochemical Technology.

Institutional Review Board Statement: Not applicable.

Informed Consent Statement: Not applicable.

Data Availability Statement: Data are contained within the article.

Acknowledgments: The team of authors express their gratitude to the reviewers and editor for valuable recommendations that have been taken into account to improve significantly the quality of this paper. The research has been carried out in the framework of the implementation of projects "Synthesis, Optimization, and parametrization of the innovative technologies of the development of coal and gas deposit resources" (0119U000248) and "Geotechnological basics of the formation of energochemical complexes of coal-mining regions" (0120U102084).

Conflicts of Interest: The authors declare no conflict of interest.

References

1. Kulikov, M.M.; Dulin, A.N.; Dulin, R.A. Prospects for Forming a Cluster for Deep Coal Processing in the Rostov Region. *Smart Innov. Syst. Technol.* **2019**, 90–98. [CrossRef]
2. Sekerin, V.; Dudin, M.; Gorokhova, A.; Bank, S.; Bank, O. Mineral resources and national economic security: Current features. *Min. Miner. Depos.* **2019**, *13*, 72–79. [CrossRef]
3. Lishtvan, I.I.; Dudarchik, V.M.; Kraiko, V.M. Prospects for the deep processing of solid fossil fuels in Belarus. *Solid Fuel Chem.* **2017**, *51*, 267–272. [CrossRef]
4. Tolmachev, O.; Urunov, A.; Muminova, S.; Dvoichenkova, G.; Davydov, I. Review of unconventional hydrocarbon resources:production technologies and opportunities for development. *Min. Miner. Depos.* **2020**, *14*, 113–121. [CrossRef]
5. Bazaluk, O.; Havrysh, V.; Nitsenko, V.; Baležentis, T.; Streimikiene, D.; Tarkhanova, E.A. Assessment of Green Methanol Production Potential and Related Economic and Environmental Benefits: The Case of China. *Energies* **2020**, *13*, 3113. [CrossRef]
6. Svyrydenko, D.; Mozgin, W. The Soft Power of the State as a Dialectic of Contemporary Dependencies in the International Arena. *Ukrainian Policymaker* **2019**, *5*, 89–97. [CrossRef] [PubMed]
7. Sai, K.; Malanchuk, Z.; Petlovanyi, M.; Saik, P.; Lozynskyi, V. Research of Thermodynamic Conditions for Gas Hydrates Formation from Methane in the Coal Mines. *Solid State Phenom.* **2019**, *291*, 155–172. [CrossRef]
8. Krichevskiy, S. Evolution of technologies, "green" development and grounds of the general theory of technologies. *Philos. Cosmol.* **2015**, *14*, 120–139.
9. Haiko, H.; Saik, P.; Lozynskyi, V. The Philosophy of Mining: Historical Aspect and Future Prospect. *Philos. Cosmol.* **2019**, *22*, 76–90. [CrossRef]
10. Fyk, M.; Biletskyi, V.; Abbood, M.; Al-Sultan, M.; Abbood, M.; Abdullatif, H.; Shapchenko, Y. Modeling of the lifting of a heat transfer agent in a geothermal well of a gas condensate deposit. *Min. Miner. Depos.* **2020**, *14*, 66–74. [CrossRef]
11. Bazaluk, O.; Slabyi, O.; Vekeryk, V.; Velychkovych, A.; Ropyak, L.; Lozynskyi, V. A Technology of Hydrocarbon Fluid Production Intensification by Productive Stratum Drainage Zone Reaming. *Energies* **2021**, *14*, 3514. [CrossRef]
12. Chen, C.; Horio, M.; Kojima, T. Numerical simulation of entrained flow coal gasifiers. Part I: Modeling of coal gasification in an entrained flow gasifier. *Chem. Eng. Sci.* **2000**, *55*, 3861–3874. [CrossRef]
13. Goncharenko, L.; Ryzhakova, A.; Sedova, N.; Efimov, I.; Akulinin, F. Survey of the world practice of implementing energy-efficient technologies in terms of mining enterprises. *Min. Miner. Depos.* **2019**, *13*, 63–71. [CrossRef]
14. Duan, T.; Lu, C.; Xiong, S.; Fu, Z.; Zhang, B. Evaluation method of the energy conversion efficiency of coal gasification and related applications. *Int. J. Energy Res.* **2015**, *40*, 168–180. [CrossRef]

15. Bhutto, A.W.; Bazmi, A.A.; Zahedi, G. Underground coal gasification: From fundamentals to applications. *Prog. Energy Combust. Sci.* **2013**, *39*, 189–214. [CrossRef]
16. Akbarzadeh Kasani, H.; Chalaturnyk, R.J. Coupled reservoir and geomechanical simulation for a deep underground coal gasification project. *J. Nat. Gas Sci. Eng.* **2017**, *37*, 487–501. [CrossRef]
17. Nakaten, N.; Kempka, T. Techno-Economic Comparison of Onshore and Offshore Underground Coal Gasification End-Product Competitiveness. *Energies* **2019**, *12*, 3252. [CrossRef]
18. Liu, H.; Liu, S. Life cycle energy consumption and GHG emissions of hydrogen production from underground coal gasification in comparison with surface coal gasification. *Int. J. Hydrogen Energy* **2021**, *46*, 9630–9643. [CrossRef]
19. Shustov, O.; Pavlychenko, A.; Bondarenko, A.; Bielov, O.; Borysovska, O.; Abdiev, A. Substantiation into Parameters of Carbon Fuel Production Technology from Brown Coal. *Mater. Sci. Forum* **2021**. Preprint.
20. Shafirovich, E.; Varma, A. Underground Coal Gasification: A Brief Review of Current Status. *Ind. Eng. Chem. Res.* **2009**, *48*, 7865–7875. [CrossRef]
21. Friedmann, S.J.; Upadhye, R.; Kong, F.-M. Prospects for underground coal gasification in carbon-constrained world. *Energy Procedia* **2009**, *1*, 4551–4557. [CrossRef]
22. Kempka, T.; Plötz, M.-L.; Schlüter, R.; Hamann, J.; Deowan, S.A.; Azzam, R. Carbon dioxide utilisation for carbamide production by application of the coupled UCG-urea process. *Energy Procedia* **2011**, *4*, 2200–2205. [CrossRef]
23. Buzylo, V.; Pavlychenko, A.; Savelieva, T.; Borysovska, O. Ecological aspects of managing the stressed-deformed state of the mountain massif during the development of multiple coal layers. *E3S Web Conf.* **2018**, *60*, 00013. [CrossRef]
24. Medunić, G.; Mondol, D.; Rađenović, A.; Nazir, S. Review of the latest research on coal, environment, and clean technologies. *Rud. Geol. Naft. Zb.* **2018**, *33*, 13–21. [CrossRef]
25. Novak, K.; Malvić, T.; Simon, K. Increased hydrocarbon recovery and CO2 management, a Croatian example. *Environ. Earth Sci.* **2012**, *68*, 1187–1197. [CrossRef]
26. Popovych, V.; Telak, J.; Telak, O.; Malovanyy, M.; Yakovchuk, R.; Popovych, N. Migration of Hazardous Components of Municipal Landfill Leachates into the Environment. *J. Ecol. Eng.* **2020**, *21*, 52–62. [CrossRef]
27. Menshov, O.; Sukhorada, A.; Homenko, R.; Kruglov, O. Ultradetailed Environmental Magnetic Investigations in Ukraine. *Near Surf. Geosci.* **2012**, cp-306-00099. [CrossRef]
28. Saik, P.; Petlevanyi, M.; Lozynskyi, V.; Sai, K.; Merzlikin, A. Innovative approach to the integrated use of energy resources of underground coal gasification. *Solid State Phenom.* **2018**, *277*, 221–231. [CrossRef]
29. Bazaluk, O.; Sai, K.; Lozynskyi, V.; Petlovanyi, M.; Saik, P. Research into Dissociation Zones of Gas Hydrate Deposits with a Heterogeneous Structure in the Black Sea. *Energies* **2021**, *14*, 1345. [CrossRef]
30. Sarhosis, V.; Kapusta, K.; Lavis, S. Underground coal gasification (UCG) in Europe: Field trials, laboratory experiments, and EU-funded projects. *Undergr. Coal Gasif. Combust.* **2018**, 129–171. [CrossRef]
31. Blinderman, M.S.; Klimenko, A.Y. Introduction to underground coal gasification and combustion. *Undergr. Coal Gasif. Combust.* **2018**, 1–8. [CrossRef]
32. Maev, S.; Blinderman, M.S.; Gruber, G.P. Underground coal gasification (UCG) to products: Designs, efficiencies, and economics. *Undergr. Coal Gasif. Combust.* **2018**, 435–468. [CrossRef]
33. Lozynskyi, V.; Medianyk, V.; Saik, P.; Rysbekov, K.; Demydov, M. Multivariance solutions for designing new levels of coal mines. *Rud. Geol. Naft. Zb.* **2020**, *35*, 23–32. [CrossRef]
34. Motovilov, I.Y.; Telkov, S.A.; Barmenshinova, M.B.; Nurmanova, A.N. Examination of the preliminary gravity dressing influence on the Shalkiya deposit complex ore. *Non Ferr. Met.* **2019**, *47*, 3–8. [CrossRef]
35. Gorova, A.; Pavlychenko, A.; Borysovs'ka, O. The study of ecological state of waste disposal areas of energy and mining companies. *Annu. Sci. Tech. Colletion* **2013**, 169–172. [CrossRef]
36. Nurpeisova, M.B.; Kurmanbaev, O.S. Laws of devolopment of geomechanical processes in the rock mass maykain mine. *News Natl. Acad. Sci. Repub. Kazakhstan Ser. Geol. Tech. Sci.* **2016**, *6*, 109–115.
37. Dychkovskyi, R.; Shavarskyi, I.; Saik, P.; Lozynskyi, V.; Falshtynskyi, V.; Cabana, E. Research into stress-strain state of the rock mass condition in the process of the operation of double-unit longwalls. *Min. Miner. Depos.* **2020**, *14*, 85–94. [CrossRef]
38. Babets, D.V.; Sdvyzhkova, O.O.; Larionov, M.H.; Tereshchuk, R.M. Estimation of rock mass stability based on probability approach and rating systems. *Nauk. Visnyk Natsionalnoho Hirnychoho Universytetu* **2017**, *2*, 58–64.
39. Petlovanyi, M.; Lozynskyi, V.; Zubko, S.; Saik, P.; Sai, K. The influence of geology and ore deposit occurrence conditions on dilution indicators of extracted reserves. *Rud. Geol. Naft. Zb.* **2019**, *34*, 83–91. [CrossRef]
40. Zeynullin, A.A.; Abeuov, E.A.; Demin, V.F.; Aliev, S.B.; Kaynazarova, A.S.; Kaynazarov, A.S. Estimation of ways to maintain mining works based on the application of anchor anchoring in the mines of the Karaganda coal basin. *Ugol* **2021**, *2*, 4–9. [CrossRef]
41. Kyrgizbayeva, G.; Nurpeisov, M.; Sarybayev, O. The monitoring of earth surface displacements during the subsoil development. *New Dev. Min. Eng.* **2015**, 161–167. [CrossRef]
42. Kolokolov, O.V.; Tabachenko, M.M.; Eyshinskiy, O.M.; Kuznetsov, V.G.; Kablanov, A.I.; Mikenberg, O.A. *Teoriya I Praktika Termohimichnoii Tehnologiii Vidobutku ta Pererobki Vugillya*; NMA: Dnipro, Ukraine, 2000; 281p.
43. Petlovanyi, M.; Lozynskyi, V.; Saik, P.; Sai, K. Predicting the producing well stability in the place of its curving at the underground coal seams gasification. *E3S Web Conf.* **2019**, *123*, 01019. [CrossRef]

44. Dubiński, J.; Turek, M. Basic Aspects of Productivity of Underground Coal Gasification Process. *Arch. Min. Sci.* **2015**, *60*, 443–453. [CrossRef]
45. Zou, C.; Chen, Y.; Kong, L.; Sun, F.; Chen, S.; Dong, Z. Underground coal gasification and its strategic significance to the development of natural gas industry in China. *Pet. Explor. Dev.* **2019**, *46*, 205–215. [CrossRef]
46. Imran, M.; Kumar, D.; Kumar, N.; Qayyum, A.; Saeed, A.; Bhatti, M.S. Environmental concerns of underground coal gasification. *Renew. Sustain. Energy Rev.* **2014**, *31*, 600–610. [CrossRef]
47. Mao, F. Underground coal gasification (UCG): A new trend of supply-side economics of fossil fuels. *Nat. Gas Ind. B* **2016**, *3*, 312–322. [CrossRef]
48. Sobczyk, E.J.; Wota, A.; Kopacz, M.; Frączek, J. Clean Coal Technologies–a chance for Poland's energy security. Decision-making using AHP with Benefits, Opportunities, Costs and Risk Analysis. *Gospod. Surowcami Miner.* **2017**, *33*, 27–48. [CrossRef]
49. Yang, L. Clean coal technology—Study on the pilot project experiment of underground coal gasification. *Energy* **2003**, *28*, 1445–1460. [CrossRef]
50. Prabu, V.; Jayanti, S. Integration of underground coal gasification with a solid oxide fuel cell system for clean coal utilization. *Int. J. Hydrogen Energy* **2012**, *37*, 1677–1688. [CrossRef]
51. Khadse, A.; Qayyumi, M.; Mahajani, S.; Aghalayam, P. Underground coal gasification: A new clean coal utilization technique for India. *Energy* **2007**, *32*, 2061–2071. [CrossRef]
52. Xie, J.; Xin, L.; Hu, X.; Cheng, W.; Liu, W.; Wang, Z. Technical application of safety and cleaner production technology by underground coal gasification in China. *J. Clean. Prod.* **2020**, *250*, 119487. [CrossRef]
53. Dychkovskyi, R.; Vladyko, O.; Maltsev, D.; Cabana, E.C. Some aspects of the compatibility of mineral mining technologies. *Rud. Geol. Naft. Zb.* **2018**, *33*, 73–82. [CrossRef]
54. Gayko, G.; Kasyanov, V. Utilizing Thermal Power Potential of Coal by Underground Burning (Gasification) of Thin Coal Layers. *Int. Min. Forum* **2007**, 97–101. [CrossRef]
55. Hu, Z.; Peng, Y.; Sun, F.; Chen, S.; Zhou, Y. Thermodynamic equilibrium simulation on the synthesis gas composition in the context of underground coal gasification. *Fuel* **2021**, *293*, 120462. [CrossRef]
56. Małkowski, P.; Niedbalski, Z.; Hydzik-Wiśniewska, J. The Change of Structural and Thermal Properties of Rocks Exposed to High Temperatures in the Vicinity of Designed Geo-Reactor. *Arch. Min. Sci.* **2013**, *58*, 465–480.
57. Wang, J.; Wang, Z.; Xin, L.; Xu, Z.; Gui, J.; Lu, X. Temperature field distribution and parametric study in underground coal gasification stope. *Int. J. Therm. Sci.* **2017**, *111*, 66–77. [CrossRef]
58. Kapusta, K.; Stańczyk, K.; Wiatowski, M.; Chećko, J. Environmental aspects of a field-scale underground coal gasification trial in a shallow coal seam at the Experimental Mine Barbara in Poland. *Fuel* **2013**, *113*, 196–208. [CrossRef]
59. Wiatowski, M.; Kapusta, K. Evolution of tar compounds in raw gas from a pilot-scale underground coal gasification (UCG) trial at Wieczorek mine in Poland. *Fuel* **2020**, *276*, 118070. [CrossRef]
60. Laciak, M.; Kostúr, K.; Durdán, M.; Kačur, J.; Flegner, P. The analysis of the underground coal gasification in experimental equipment. *Energy* **2016**, *114*, 332–343. [CrossRef]
61. Mocek, P.; Pieszczek, M.; Świądrowski, J.; Kapusta, K.; Wiatowski, M.; Stańczyk, K. Pilot-scale underground coal gasification (UCG) experiment in an operating Mine "Wieczorek" in Poland. *Energy* **2016**, *111*, 313–321. [CrossRef]
62. Su, F.Q.; Itakura, K.; Hamanaka, A.; Deguchi, G.; Sato, K.; Kodama, J. Ex Situ UCG Model Experiments with Oxygen Enriched Air in an Artificial Coal Seam. *Key Eng. Mater.* **2017**, *737*, 379–384. [CrossRef]
63. Doucet, D.; Perkins, G.; Ulbrich, A.; du Toit, E. Production of power using underground coal gasification. *Energy Sources Part A Recovery Util. Environ. Eff.* **2016**, *38*, 3653–3660. [CrossRef]
64. Pivnyak, G.; Dychkovskyi, R.; Bobyliov, O.; Cabana, E.C.; Smoliński, A. Mathematical and Geomechanical Model in Physical and Chemical Processes of Underground Coal Gasification. *Solid State Phenom.* **2018**, *277*, 1–16. [CrossRef]
65. Stańczyk, K.; Howaniec, N.; Smoliński, A.; Świądrowski, J.; Kapusta, K.; Wiatowski, M.; Rogut, J. Gasification of lignite and hard coal with air and oxygen enriched air in a pilot scale ex situ reactor for underground gasification. *Fuel* **2011**, *90*, 1953–1962. [CrossRef]
66. Seifi, M.; Chen, Z.; Abedi, J. Numerical simulation of underground coal gasification using the CRIP method. *Can. J. Chem. Eng.* **2011**, *89*, 1528–1535. [CrossRef]
67. Li, H.; Zha, J.; Guo, G.; Zheng, N.; Gong, Y. Improvement of resource recovery rate for underground coal gasification through the gasifier size management. *J. Clean. Prod.* **2020**, *259*, 120911. [CrossRef]
68. Uwaoma, R.C.; Strydom, C.A.; Matjie, R.H.; Bunt, J.R.; van Dyk, J. The influence of the roof and floor geological structures on the ash composition produced from coal at UCG temperatures. *Int. J. Coal Prep. Util.* **2018**, *40*, 247–265. [CrossRef]
69. Falshtynskyi, V.; Saik, P.; Lozynskyi, V.; Dychkovskyi, R.; Petlovanyi, M. Innovative aspects of underground coal gasification technology in mine conditions. *Min. Miner. Depos.* **2018**, *12*, 68–75. [CrossRef]
70. Li, H.; Guo, G.; Zheng, N. Influence of coal types on overlying strata movement and deformation in underground coal gasification without shaft and prediction method of surface subsidence. *Process Saf. Environ. Prot.* **2018**, *120*, 302–312. [CrossRef]
71. An, N.; Zagorščak, R.; Thomas, H.R.; Gao, W. A numerical investigation into the environmental impact of underground coal gasification technology based on a coupled thermal-hydro-chemical model. *J. Clean. Prod.* **2021**, *290*, 125181. [CrossRef]
72. Róg, L. Vitrinite reflectance as a measure of the range of influence of the temperature of a georeactor on rock mass during underground coal gasification. *Fuel* **2018**, *224*, 94–100. [CrossRef]

73. Kostúr, K.; Laciak, M.; Durdan, M. Some Influences of Underground Coal Gasification on the Environment. *Sustainability* **2018**, *10*, 1512. [CrossRef]
74. Dvornikova, E.V. Environmental performance of underground coal gasification. *Undergr. Coal Gasif. Combust.* **2018**, 363–399. [CrossRef]
75. Shahbazi, M.; Najafi, M.; Marji, M.F. On the mitigating environmental aspects of a vertical well in underground coal gasification method. *Mitig. Adapt. Strateg. Glob. Chang.* **2019**, *24*, 373–398. [CrossRef]
76. Burchart-Korol, D.; Krawczyk, P.; Czaplicka-Kolarz, K.; Smoliński, A. Eco-efficiency of underground coal gasification (UCG) for electricity production. *Fuel* **2016**, *173*, 239–246. [CrossRef]
77. Wang, G.X.; Wang, Z.T.; Feng, B.; Rudolph, V.; Jiao, J.L. Semi-industrial tests on enhanced underground coal gasification at Zhong-Liang-Shan coal mine. *Asia Pac. J. Chem. Eng.* **2009**, *45*, 771–779. [CrossRef]
78. Saik, P.B.; Dychkovskyi, R.O.; Lozynskyi, V.H.; Malanchuk, Z.R.; Malanchuk, Y.Z. Revisiting the underground gasification of coal reserves from contiguous seams. *Nauk. Visnyk Natsionalnoho Hirnychoho Universytetu* **2016**, *6*, 60–66.
79. Zhu, L.; Wang, F.; Zhang, Z. Thermodynamic evaluation of a conceptual process for coal gasification coupled with chemical looping air separation. *Chem. Eng. Process. Process Intensif.* **2016**, *106*, 33–41. [CrossRef]
80. Koroviaka, Y.; Pinka, J.; Tymchenko, S.; Rastsvietaiev, V.; Astakhov, V.; Dmytruk, O. Elaborating a scheme for mine methane capturing while developing coal gas seams. *Rozrobka Rodovyshch* **2020**, *14*, 21–27. [CrossRef]
81. Yang, L. Coal Properties and System Operating Parameters for Underground Coal Gasification. *Energy Sources Part A Recovery Util. Environ. Eff.* **2008**, *30*, 516–528. [CrossRef]
82. Lazarenko, S.N.; Kochetkov, V.N. The underground coal gasification is the technology which answers to conditions of sustainable development of coal region. *Mine Plan. Equip. Sel.* **2020**, 167–168. [CrossRef]
83. Falshtynskyi, V.; Lozynskyi, V.; Saik, P.; Dychkovskyi, R.; Tabachenko, M. Substantiating parameters of stratification cavities formation in the roof rocks during underground coal gasification. *Min. Miner. Depos.* **2016**, *10*, 16–24. [CrossRef]
84. Majkherchik, T.; Gajko, G.I.; Malkowski, P. Deformation process around a heading investigation when front of longwall face advancing. *Ugol* **2002**, *11*, 27–29.
85. Falshtynskyi, V.S.; Dychkovskyi, R.O.; Lozynskyi, V.G.; Saik, P.B. Determination of the Technological Parameters of Borehole Underground Coal Gasification for Thin Coal Seams. *J. Sustain. Min.* **2013**, *12*, 8–16. [CrossRef]
86. Prusek, S.; Lubosik, Z.; Rajwa, S.; Walentek, A.; Wrana, A. Geotechnical monitoring of rock mass and support behaviour around the UCG georeactor: Two case studies in Polish coal mining industry. In Proceedings of the International Conference on Ground Control in Mining, Hangzhou, China, 27–29 October 2017; pp. 321–328.
87. Bondarenko, V.; Dychkovskiy, R.; Falshtynskiy, V. Synthetic Stowing of Rockmass at Borehole Underground Coal Gasification (BUCG). *Deep Min. Chall.* **2009**, 169–177. [CrossRef]
88. Buzylo, V.; Bondarenko, V.; Dychkovskiy, R.; Falshtynskiy, V. Parameters of Injection Fill above an Underground Gas Generator. Technical, Technological and Economical Aspects of Thin-Seams Coal Mining. *Int. Min. Forum* **2007**, 89–95. [CrossRef]
89. Falshtynskyi, V.; Dychkovskyi, R.; Saik, P.; Lozynskyi, V. Someaspectsoftechnologicalprocessescontrolofanin-situgasifierduringcoalseamgasi In *Progressive Technologies of Coal, Coalbed Methane, and Ores Mining*; CRC Press: London, UK, 2014; pp. 109–112. [CrossRef]
90. Saik, P.; Lozynskyi, V.; Chemeriachko, Y.; Cabana, E. Basics of the approach formation to substantiate the temperature field distribution during experimental research on the coal gasification processes. *E3S Web Conf.* **2020**, *202*, 01037. [CrossRef]
91. Falshtynskyi, V.; Dychkovskyi, V.; Lozynskyi, V.; Saik, P. New method for justification the technological parameters of coal gasification in the test setting. *Geomech. Process. Dur. Undergr. Min. Proc. Sch. Undergr. Min.* **2012**, 201–208. [CrossRef]
92. Durdán, M.; Terpák, J.; Kačur, J.; Laciak, M.; Flegner, P. Modeling of material balance from the experimental UCG. *Acta Polytech.* **2020**, *60*, 391–399. [CrossRef]
93. Lozynskyi, V.; Dychkovskyi, R.; Saik, P.; Falshtynskyi, V. Coal Seam Gasification in Faulting Zones (Heat and Mass Balance Study). *Solid State Phenom.* **2018**, *277*, 66–79. [CrossRef]
94. Lozynskyi, V.; Saik, P.; Petlovanyi, M.; Sai, K.; Malanchuk, Z.; Malanchyk, Y. Substantiation into mass and heat balance for underground coal gasification in faulting zones. *Inz. Miner.* **2018**, *19*, 289–300. [CrossRef]
95. Uciechowska-Grakowicz, A.; Strzelecki, T. Application of the thermoporoelasticity model in numerical modelling of underground coal gasification influence on the surrounding medium. *Studia Geotech. Et Mech.* **2021**. in Press. [CrossRef]
96. Janoszek, T.; Stańczyk, K.; Smoliński, A. Modelling Test of Autothermal Gasification Process Using CFD. *Arch. Min. Sci.* **2017**, *62*, 253–268. [CrossRef]
97. Lozynskyi, V.H. *Substantiation into Parameters of Borehole Underground Coal Gasification Technology in the Faulting Zone*; National Mining University (Dnipro University of Technology): Dnipro, Ukraine, 2015; pp. 1–156.
98. Pivnyak, G. Conditions of Suitability of Coal Seams for Underground Coal Gasification. *Key Eng. Mater.* **2020**, *844*, 38–48. [CrossRef]
99. Savostianov, O.V. Matematychna model napryzhenno-deformovanoho stanu porodnoho masyvy pry pidzemnii hazyfikacii tverdoho palyva. *Collect. Res. Pap. Natl. Min. Univ.* **2008**, *30*, 44–56.
100. Falshtynskyi, V.; Dychkovskyi, R.; Illiashov, M. Engineering support of BUCG process in Solenovsk coal deposits. *Tech. Geoinf. Syst. Min.* **2011**, 47–56. [CrossRef]
101. Hongtao, L.; Feng, C.; Xia, P.; Kai, Y.; Shuqin, L. Method of oxygen-enriched two-stage underground coal gasification. *Min. Sci. Technol.* **2011**, *21*, 191–196. [CrossRef]

102. Yang, L.H.; Zhang, X.; Liu, S. Underground Coal Gasification Using Oxygen and Steam. *Energy Sources Part A Recovery Util. Environ. Eff.* **2009**, *31*, 1883–1892. [CrossRef]
103. Eftekhari, A.A.; Wolf, K.H.; Rogut, J.; Bruining, H. Energy and exergy analysis of alternating injection of oxygen and steam in the low emission underground gasification of deep thin coal. *Appl. Energy* **2017**, *208*, 62–71. [CrossRef]
104. Liu, S.; Liang, J.; Chang, J.; Yang, Z.; Yu, L. UCG model test of Huating coal with oxygen-steam as gasification agent. *J. Southeast Univ.* **2003**, *33*, 355–358.
105. Sadasivam, S.; Zagorščak, R.; Thomas, H.R.; Kapusta, K.; Stańczyk, K. Experimental study of methane-oriented gasification of semi-anthracite and bituminous coals using oxygen and steam in the context of underground coal gasification (UCG): Effects of pressure, temperature, gasification reactant supply rates and coal rank. *Fuel* **2020**, *268*, 117330. [CrossRef]
106. Marcourt, M.; Paquay, V.; Piel, A.; Pirard, J.-P. Coal gasification at pressure by mixtures of carbon dioxide and oxygen. *Fuel* **1983**, *62*, 823–828. [CrossRef]
107. Chodankar, C.R.; Feng, B.; Ran, J.; Klimenko, A.Y. Kinetic study of the gasification of chinchilla coal char in carbon dioxide. *Annu. Int. Pittsburgh Coal Conf.* **2007**, *3*, 1280–1291.
108. Jayaraman, K.; Gokalp, I. Effect of char generation method on steam, CO_2 and blended mixture gasification of high ash Turkish coals. *Fuel* **2015**, *153*, 320–327. [CrossRef]
109. Liu, H.; Chen, F.; Wang, Y.; Liu, G.; Yao, H.; Liu, S. Experimental Study of Reverse Underground Coal Gasification. *Energies* **2018**, *11*, 2949. [CrossRef]
110. Cui, Y.; Liang, J.; Wang, Z.; Zhang, X.; Fan, C.; Liang, D.; Wang, X. Forward and reverse combustion gasification of coal with production of high-quality syngas in a simulated pilot system for in situ gasification. *Appl. Energy* **2014**, *131*, 9–19. [CrossRef]
111. Cui, Y.; Liang, J.; Wang, Z.; Zhang, X.; Fan, C.; Wang, X. Experimental forward and reverse in situ combustion gasification of lignite with production of hydrogen-rich syngas. *Int. J. Coal Sci. Technol.* **2014**, *1*, 70–80. [CrossRef]
112. Elahi, S.M.; Nassir, M.; Chen, Z. Effect of various coal constitutive models on coupled thermo-mechanical modeling of underground coal gasification. *J. Pet. Sci. Eng.* **2017**, *154*, 469–478. [CrossRef]
113. Perkins, G. Underground coal gasification–Part I: Field demonstrations and process performance. *Prog. Energy Combust. Sci.* **2018**, *67*, 158–187. [CrossRef]
114. Kashyap, S.; Vairakannu, P. Movable injection point–based syngas production in the context of underground coal gasification. *Int. J. Energy Res.* **2020**, *44*, 3574–3586. [CrossRef]
115. Yang, L. Study on the Method of Two-Phase Underground Coal Gasification with Unfixed Pumping Points. *Energy Sour.* **2003**, *25*, 917–930. [CrossRef]
116. Chen, Z.; Zhu, F.; Zhang, Y.; Lv, W.; Zhang, Z. Development of gasification agent injection tool for underground coal gasification. *E3S Web Conf.* **2021**, *267*, 02056. [CrossRef]

energies

MDPI

Article

Regression Models Utilization to the Underground Temperature Determination at Coal Energy Conversion

Milan Durdán *,†, Marta Benková *,†, Marek Laciak †, Ján Kačur † and Patrik Flegner †

Institute of Control and Informatization of Production Processes, Faculty of Mining, Ecology, Process Control and Geotechnologies, Technical University of Košice, Němcovej 3, 042 00 Košice, Slovakia; marek.laciak@tuke.sk (M.L.); jan.kacur@tuke.sk (J.K.); patrik.flegner@tuke.sk (P.F.)
* Correspondence: milan.durdan@tuke.sk (M.D.); marta.benkova@tuke.sk (M.B.); Tel.: +421-55-602-5178 (M.D.)
† These authors contributed equally to this work.

Abstract: The underground coal gasification represents a technology capable of obtaining synthetic coal gas from hard-reached coal deposits and coal beds with tectonic faults. This technology is also less expensive than conventional coal mining. The cavity is formed in the coal seam by converting coal to synthetic gas during the underground coal gasification process. The cavity growth rate and the gasification queue's moving velocity are affected by controllable variables, i.e., the operation pressure, the gasification agent, and the laboratory coal seam geometry. These variables can be continuously measured by standard measuring devices and techniques as opposed to the underground temperature. This paper researches the possibility of the regression models utilization for temperature data prediction for this reason. Several regression models were proposed that were differed in their structures, i.e., the number and type of selected controllable variables as independent variables. The goal was to find such a regression model structure, where the underground temperature is predicted with the greatest possible accuracy. The regression model structures' proposal was realized on data obtained from two laboratory measurements realized in the ex situ reactor. The obtained temperature data can be used for visualization of the cavity growth in the gasified coal seam.

Keywords: underground coal gasification; measurement; temperature; regression; model; analyses; cavity

Citation: Durdán, M.; Benková, M.; Laciak, M.; Kačur, J.; Flegner, P. Regression Models Utilization to the Underground Temperature Determination at Coal Energy Conversion. *Energies* **2021**, *14*, 5444. https://doi.org/10.3390/en14175444

Academic Editors: Adam Smoliński and Sarma V. Pisupati

Received: 20 July 2021
Accepted: 26 August 2021
Published: 1 September 2021

1. Introduction

The underground coal gasification process (i.e., the UCG process) is a constantly evolving technology and provides an alternative to conventional coal mining. This technology transforms coal into high-calorific gas (i.e., syngas), and for coal mines located in great depths is especially effective. In implementing this technology, at least one injection and one production well must be drilled from the earth's surfaces in an area where the coal seam is located. A gasification agent (i.e., the ratio of air, oxygen, and water vapor) is injected through the injection well. This gasification agent will ensure the chemical reactions occur. These chemical reactions are required for the syngas creating. Subsequently, Syngas is extracted through the production well and subsequently cleaned and stored (see Figure 1).

The UCG reactor can be divided into three basic zones in terms of the chemical reactions that occur. Chemical reactions to increase the coal seam temperature are taking place in the oxidation zone (i.e., at a temperature above 900 °C). In the reduction zone (i.e., at a temperature between 550–900 °C), chemical reactions transform coal into syngas (i.e., a mixture of CO, CO_2, CH_4, H_2, etc.). The pyrolysis and drying process of the coal seam takes place in the drying and pyrolysis zone (i.e., at a temperature between 220–550 °C). All these processes take place at a desired coal seam temperature. It is important to know the temperature distribution in the coal seam for this reason. The temperature information can determine the distribution of the individual zones in the coal seam and set the appropriate gasification agent mixture to increase the coal seam temperature or create syngas. In addition, the cavity formation process in coal seam can be observed by seeing behaviors of temperatures [1,2].

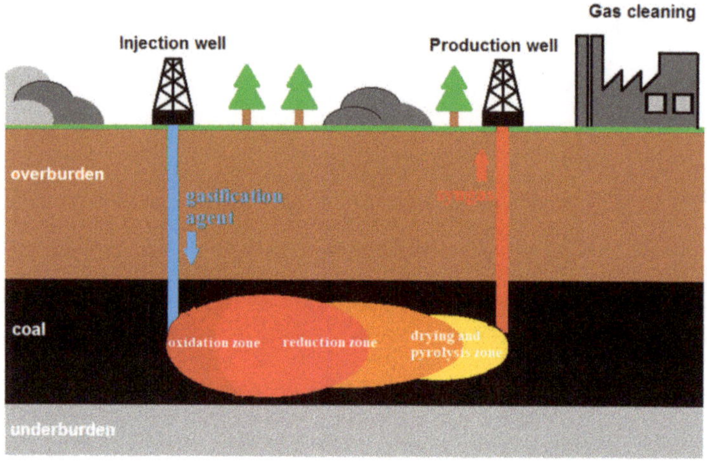

Figure 1. The scheme of the UCG process (Source: own elaboration).

Model in the form of a risk graph and a risk matrix was used for risk analysis and hazard identification in the UCG process by investigating health risks and impacts in terms of their influence on the environment [3]. The proposed measures could lead to the reduction of risks and impacts on an acceptable level. The UCG process state prediction was realized by a dual-source long short-term memory (i.e., LSTM) prediction model [4]. This model was compared with the Support Vector Machine (i.e., SVM) and Back Propagation Neural Network (i.e., BPNN) prediction model. The results showed that the predicting trends accuracy reaches 90.99%. The prediction of syngas composition was realized by a thermochemical equilibrium model of the UCG process [5]. This model considers the effect of the drying process and is based on the water–gas shift reaction and gasification reactions. The results showed the positive impact of the steam addition into the gasification agent to increase the hydrogen and carbon monoxide content in syngas and increase the calorific value. A stoichiometric equilibrium model has been used to estimate the equilibrium composition of the produced gas [6]. This model is based on the Gibbs function's minimization and was used to simulate the relevant thermochemical coal conversion processes. Verification of described model showed that the produced gas composition was in a good agreement under different operating conditions. A three-dimensional numerical model has been used to simulate an ex situ allothermal coal gasification experiment [7]. The deviations between the simulated composition of produced gas and experimental data were from 10% (e.g., H_2) to less than 50% (e.g., CH_4) at some coal samples. The prediction of the shape and volume of the underground cavity over time was realized by the model based on a series of equations, the cavity pressure, and temperature information [8]. The simulations were realized by COMSOL software, and the results in the form percentage of the product gas components (i.e., CO, CO_2, CH_4, and H_2) showed a good comparison. A review of various gasification UCG models for predicting the cavity growth and the product gas recovery where the temperature value (i.e., the cavity temperature, solid-phase temperature, the gasification agent temperature, and the produced gas temperature) has the critical role was described [9]. The results of this review showed that the packed bed models are applicable for highly permeable porous media, the channel models overcome the limitation of the packed bed models in regards to calculating the cavity shape and size, and the coal slab models describe the process by the movement of the various defined regions in the coal slab. An empirical model based on the non-linear multivariable regression method also realized the prediction of cavity growth [10]. Nine possible independent variables as moisture, operating pressure, seam thickness, seam depth, calorific value, permeability, volatile matter, and fixed carbon were evaluated in terms of their prediction ability during the analysis. The regression analysis excluded the coal seam thickness and fixed carbon content. The investigation of reaction zones, reaction rates, cavity formation, and syngas properties was made by a two-dimensional computational fluid dynamics model [11]. This model was verified by numerical calculation of the syngas composition and a lab-scale hydrogen experiment. A good agreement was shown between the calculated and the experimental data at every stage. A laboratory-based UCG model with advanced real-time control and monitoring was used for monitoring cavity formation, temperatures, syngas characteristics, and coal-tar [12]. Results showed the ability of the laboratory-based UCG process to forecast the sustainability and standardization before the UCG process implementation by observed of the realized experiment. It has been demonstrated a significant influence of operational pressure and coal properties on the process parameters as energy efficiency, gas composition, and methane yields by realized model experiments [13]. The oscillations of the gas production rates reflected the changes in gasification conditions and the cavity geometry. A dependence was shown of methane yields on the properties of coal and the pressure regime. The Continuous Retracting Injection Point (i.e., CRIP) Process Model and the Linked Vertical Wells (i.e., LVW) Process Model were used for modeling the rate of cavity growth and the related chemical process in the UCG process [14]. The CRIP Process Model showed the more controllable cavity growth rate in a better-regulated water influx from the surrounding strata.

Nowadays, it is of great interest in the methods area which examining temperatures distribution in the coal seam. The temperature field calculation was solved by two-dimension nonlinear unstable mathematical models and analyzing the regularity of the temperature field distribution in the gasified coal layers of the UCG reactor [15]. The laboratory model experiment in a laboratory gasifier was used to establish the accuracy of this methodology. The most significant differences between calculated results and measured temperatures were in the combustion zone (i.e., some temperature points above 20%), but differences of other measuring points are below 15%, most of which, within 10%. The analytical solution of one-dimensional unsteady heat conduction was used to study temperature distribution in burnt surrounding rocks at the UCG process [16]. The heat conduction is solved by the first and the fourth kinds of boundary conditions in this solution. The modeled case showed that the temperature influence range in burnt surrounding rock is circa 18–19 m. The two-dimensional temperature field of the UCG reactor was also solved by the heat conduction model based on the first and third kinds of boundary conditions and velocity of motion of the gasification front [17]. There was examined the influence of burned coal seam (i.e., the heat source) on surroundings rocks, including the earth surface. The modeled cases showed that the temperature changes of the surrounding rock were maximal 14 m from the boundary of the coal and overburden towards the earth's surface. A mathematical prediction model using the CFD software package (FLUENT 6.3.26) was used to predict cavity growth, temperature distribution, and coal consumption [18]. The CFD software package solved the simulation of combustion and gasification reactions on the interface between the coal seam and cavity. The error of the coal consumption prediction was less than 5%. During the underground coal gasification process, the temperature change in rock strata was identified in the numerical simulation based on the computational fluid dynamics formalism [19]. The applied software allowed the creation of coal gasification processes models at different conditions and, mainly, the process occurring beneath the ground surface. The modeled study showed temperature changes from 0.5 (i.e., 1000 °C) to 2.5 (i.e., 75 °C) meters above the gasification channel. Thermo-mechanical simulations were used to quantify the permeability changes in representative coal measure strata surrounding the UCG reactor [20]. The influence of the temperature-dependent and temperature-independent rock properties on the spatial permeability development was compared in this study. The results showed that temperature-dependent parameters are required for simulations in the close vicinity of the reactor.

Various statistical methods are used to predict values in the UCG process, including regression analysis. Regression analysis is a powerful statistical method that researchers widely use to examine the relationship between two or more variables of interest. The differential equation based on linear regression was developed to modeling the carbon dioxide emission data [21]. The penalized least-square fitting criteria were used to smooth the data. Optimizing the profile error sum of squares was used for the estimation of differential operators using functional regression. The logistic regression model from four statistical models reached the highest probability of predicting future pipe accidents [22]. It was shown that the most effective variables are the length, diameter, material, and hydraulic pressure in the pipe failure areas. The vector regression method was used for modeling the coal gross calorific value [23]. This method showed that carbon, moisture, ash, and hydrogen contents in the coal are the most effective variables for the gross calorific value modeling. The correlation of determination (R-2) for models was 0.99. Several statistical approaches solved the online coal calorific value prediction based on the flame radiation features in linear and nonlinear regression analyses [24]. The partial least squares analysis-based nonlinear regression model showed the best performance for coal calorific value prediction. The regression model is also used to calculate the pulverized coal ignition temperature [25]. The results showed that the multivariate regression method is useful for determining the ignition temperature calculation formula. Multivariable regression and artificial neural network methods were used for a wide range of coal samples from a calorific value of 10.05 to 34.80 MJ/kg [26]. The correlation coefficient values 0.77, 0.75, and 0.81 were reached by the least square mathematical method at the investigation of the relationship between inputs parameter (i.e., moisture, volatile matter, ash, total sulfur, etc.) and HGI (i.e., Hardgrove Grindability Index) in linear condition. A multivariate adaptive regression splines (i.e., MARS) approach was used for predicting the syngas temperature [27]. This proposed approach was tested in the fire prevention area of UCG processes. The effect of the coal rank to examine the composition and toxicity of water effluents was solved by statistical analysis [28]. The principal component analysis, Pearson correlation analysis, and the multiple regression statistical method were used to predict the toxicity using the values of the selected parameters. The proposed regression model had a high coefficient of determination $R^2 = 0.956$ to experimental data. The study for identifying physicochemical parameters of river water that affect the electrical conductivity and evaluate their percentage contribution was realized [29]. The correlation coefficients calculation and display of the various parameters regression equations with electrical conductivity were realized by statistical analysis in this study. It is found that total dissolved solids have the highest contribution (39.6%) while total alkalinity has the second-highest contribution (23.5%), followed by total hardness (19.9%). A multiple regression model was proposed for a real-time surface roughness prediction system [30]. The proposed models with linear correlation coefficients of 0.940 and 0.933 for predictor variables, such as feed rate, vibration amplitude average, spindle speed, and depth of cut, had a strong linear correlation with the predicted variable. The regression model had an accuracy of above 90% in predicting the surface roughness. A new data mining algorithm has been proposed to capture the non-linearity in data and also find the best subset model [31]. This proposed algorithm based on the classical least square regression framework is compared with the five nodes of the neural network method. The correlation coefficient was 0.79 in the proposed algorithm and 0.81 in the neural network method. The UCG data prediction in laboratory conditions was realized by the utilization support vector machines method [32]. This method analyzed data used for classification and regression analysis to predict the underground temperature and syngas calorific value. The results obtained from the Matlab program and its statistical toolbox showed that the most appropriate is to use the Gaussian kernel function to achieve the best prediction quality. Statistical data processing was realized to investigate the relationships between measured quantities during the atmospheric geochemical survey of contaminated soil and the environmental burden of the industrial establishment [33]. The dependency between examined values was confirmed by regressive and correlative analysis.

It is necessary to continue to develop methods that would improve the prediction of the UCG process state in its implementation process due to the specificity of this process, mainly its diversity, which is determined by different geological environments. This improvement includes the accuracy improvement of the underground temperature calculation, which leads to the achievement of the required behavior of chemical reactions, the range determination of the surrounding rock burning, the prediction of cavity growth, and the produced gas's composition determination. It could lead to an increase in the efficiency of this process by producing gas with the highest possible calorific value (i.e., obtaining the maximum amount of energy) in the process control while minimizing the negative impact on the environment. We focused on using regression analysis methods to model the temperatures of the gasified coal seam in an ex situ reactor due to the wide range of applications of regression analysis methods in the processes of extraction and processing raw material. For this goal, two experiments with the same structure of the coal model differed by the amount of gasified coal were performed. Regression models were created from the data of the first experiment and subsequently verified on the data from the second experiment. Verification of the suitability of the created regression models for their use in different conditions was performed, i.e., a different amount of gasified coal and thus also different amounts of gasification agent and a time of the experiment.

2. Experiments Methodology

The methodology of the UCG process physical modeling in experimental equipment is widely used by researchers (e.g., in [34,35]). For this reason, an experimental gasifier was designed and constructed to realize the UCG process experimental measurements. The UCG process experiments were performed in an experimental laboratory gasifier (i.e., ex situ reactor). This ex situ reactor has a length of 3000 mm and a height of 500 mm and comprises two basic parts, i.e., a vessel and lid. The ex situ reactor vessel is semi-cylindrical in shape and consists of the vessel jacket and the forehead (i.e., front and rear). The inner surface of the ex situ reactor vessel is covered 100 mm thick in the isolation, which is placed under the steel cover plate. The scheme of the experimental coal gasification system is shown in Figure 2. A fan was placed behind the reactor vessel to direct the flow of the gasification agent through the coal model.

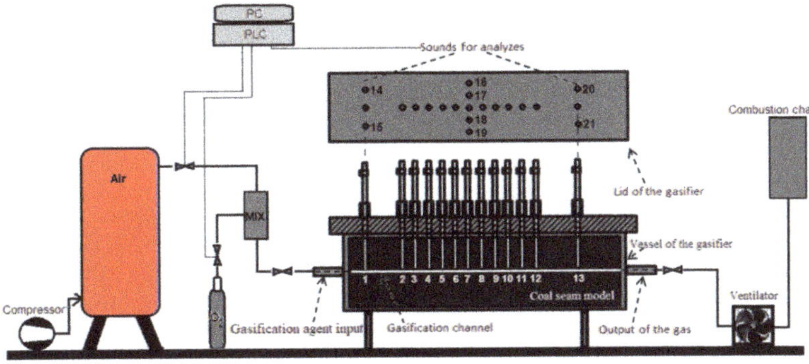

Figure 2. Scheme of the experimental coal gasification system (Source: own elaboration).

The coal seam model embedded into the ex situ reactor included the overburden, underburden, and coal blocks. This model was arranged so that the gasification agent could permeate through the whole coal seam model (i.e., the gasification channel was drilled through the entire coal seam model). The experiments were based on the regulated supply of the gasification agent (i.e., through the gasification agent input) into burning the coal seam model (i.e., embedded into the ex situ reactor) and exhaust of the syngas (i.e., through the output of the gas). The gasification agent composition was set by the ratio of the air and the oxygen (i.e., O_2). The syngas composition consisted of the following components ratio, i.e., the carbon monoxide (i.e., CO), the oxygen, the methane (i.e., CH_4), the hydrogen (i.e., H_2), etc. The syngas extraction and temperature measurement were realized by sounds placed on the ex situ reactor lid. Thermocouples measured the channel and coal temperatures, i.e., thirteen thermocouples were placed in the gasification channel (i.e., 1–13 sounds), and eight thermocouples were placed in the coal (i.e., 14–21 sounds).

Measured values, i.e., channel and coal temperatures, gasification agent and syngas composition, and their flows, were transferred from the ex situ reactor to the PC. These values were processed and shown by the monitoring system. The control of the UCG process was based on the evaluation of these values by the control algorithm. Two experiments in the described experimental ex situ reactor were realized for the regression modeling of the measured temperature values.

2.1. The First Experiment

The one layer of coal cubes with a total weight of 214 kg was embedded into the ex situ reactor. This layer had circa 30 cm the width and circa 25 cm the height. The technical analysis of these coal samples was carried in an accredited laboratory, and its results are shown in Table 1. The individual coal blocks were glued with a mixture of gudron, coal dust, and water. The cross-sectional design of the coal seam model for this experiment is shown in Figure 3. A gasification channel along the length of the whole ex situ reactor was created in the bottom third of the coal seam model height (see Figure 4a). The gasification channel had a diameter of 20 mm. The coal blocks layer was covered with a thermal insulation foil because of the prevention of heat leakage at the UCG process (see Figure 4b).

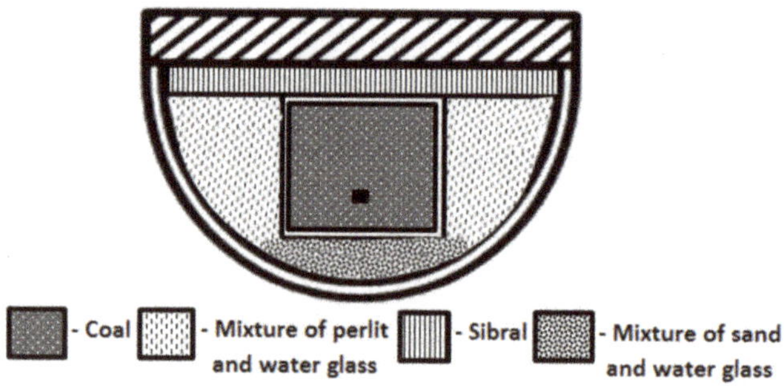

Figure 3. The cross-sectional designs of the coal model for the first experiment (Source: own elaboration based on [36]).

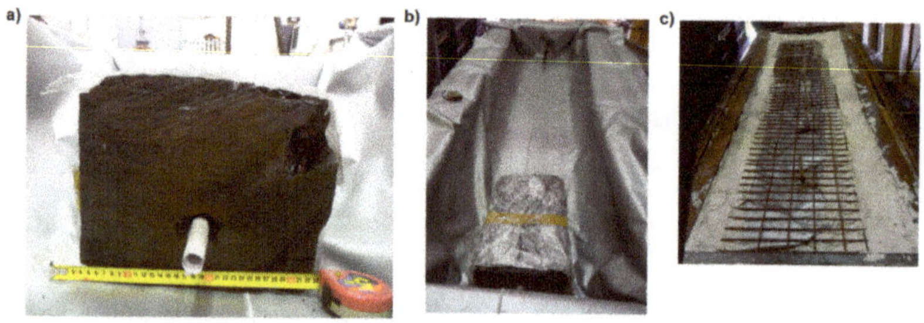

Figure 4. The construction of the coal model (**a**) gasification channel placement, (**b**) thermal insulation of coal blocks, (**c**) reinforcement of coal model top part (Source: own elaboration).

Table 1. The analysis of coal with the help of Slovak testing standards used by an accredited laboratory (Abbreviations: r—received, d—dry, daf—dry, ash-free, a—analytical, G—Gravimetry, EA—elementary analysis with heat conductive detector, K—Calorimetry, RFS—X-ray fluorescence spectrometry) (Source: own elaboration).

Parameter	Value	Uncertainty	Method	Standard
Total Moisture W_t^r (%)	38.2	5	G	PN 16.3
Ash A^d (%)	9.4	2	G	PN 16.4
Volatiles V^{daf} (%)	50	4	G	PN 16.2
Carbon C^{daf} (%)	76.5	2	EA	PN 16.7
Hydrogen H^{daf} (%)	3.95	5	EA	PN 16.7
Nitrogen N^{daf} (%)	1.48	40	EA	PN 16.7
Calorific Value Q_i^{daf} (MJ/kg)	30.2	2	K	PN 16.2
Calorific Value Q_i^d (MJ/kg)	27.4	2	K	PN 16.1
Calorific Value Q_i^r (MJ/kg)	16.0	2	K	PN 16.1
Ash A^r (%)	5.81	2	G	PN 16.4
Carbon C^r (%)	42.8	2	EA	PN 16.7
Hydrogen H^r (%)	2.21	5	EA	PN 16.7
Nitrogen N^r (%)	0.83	20	EA	PN 16.7
CaO (%)	2.37	5	RFS	PN 3.1
MgO (%)	0.46	10	RFS	PN 3.1
SiO_2 (%)	1.23	10	RFS	PN 3.1
Al_2O_3 (%)	0.74	10	RFS	PN 3.1
Fe_2O_3 (%)	1.02	10	RFS	PN 3.1
Na_2O (%)	<0.2		RFS	PN 3.1
P_2O_5 (%)	<0.02		RFS	PN 3.1
TiO_2 (%)	0.02	10	RFS	PN 3.1
K_2O (%)	0.06	10	RFS	PN 3.1
Volatiles V^r (%)	28	4	G	PN 16.2
Analytical Moisture W^a (%)	21.5	5	G	PN 16.3
Total Sulphur S_t^r (%)	1.62	15	G	PN 16.5
Sulphate Sulphur S_s^r (%)	0.17	15	G	PN 16.5
Pyritic Sulphur S_p^r (%)	0.99	15	G	PN 16.5
Organic Sulphur S_o^r (%)	1.44	15	G	PN 16.5

The thermal insulation foil separated coal blocks and a mixture of perlit and the water glass. The mixture of perlit and the water glass created isolation around the top and sides of the coal blocks. The inner bottom part of the ex situ reactor was covered with a mixture of sand and water glass. The top part of the isolation was reinforced with the steel construction because of the prevention of the isolation fall after combustion of coal blocks (see Figure 4c). The sibral plate was placed over the top part of the insulation. The mixture of perlit and the water glass, the mixture of the sand and the water glass, and sibral simulated the surrounding rock of the coal seam. The analysis shown in Figure 5 confirmed that whole coal blocks were burned.

2.2. The Second Experiment

The ex situ reactor was filled with a layer of coal cubes in a total weight of 472 kg for this experiment. The same type of coal was used as in the first experiment, i.e., coal composition is shown in Table 1. The coal seam model with the isolation around had a similar shape as is shown in Figure 3, but the gasification channel had a diameter of 40 mm. The coal blocks were cemented by a mixture of gudron, coal dust, and water (see Figure 6).

The thickness of the unburnt coal was different along the whole ex situ reactor length. In the first meter, the thickness of the unburnt coal was about 3–4 cm, in the second meter, it was around 4–6 cm (see Figure 7a), and in the third meter, it was about 6–8 cm at the edges even more, up to 16 cm as is shown in Figure 7b. There was 66 kg of unburned coal that is circa 14% from input coal.

Figure 5. The coal blocks in the experimental generator after the gasification process (Source: own elaboration.

Figure 6. The coal blocks cemented to each other by a mixture of gudron (Source: own elaboration).

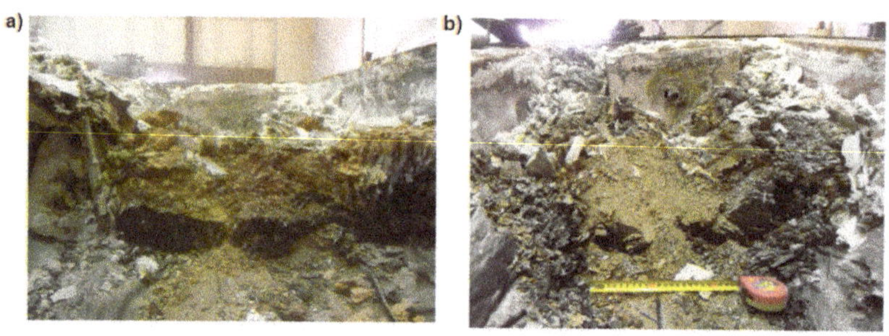

Figure 7. The unburned layer of coal (**a**) in the second meter, (**b**) in the third meter (Source: own elaboration).

3. Regression Methodology

It is often necessary to take into account that one dependent variable y is affected by several independent variables $x_1 x_2, \ldots, x_k$ [37]. We can write it in the following form:

$$y = \beta_0 + \beta_1 x_1 + \beta_2 x_2 + \ldots + \beta_k x_k + \varepsilon, \tag{1}$$

where the regression parameter β_j $j = 1, 2, \cdots, k$, expresses the assumed change in the variable y caused by the unit change of one independent variable x_j, if the other independent variables do not change; ε is a random error.

The least-squares method is most often used to estimate the regression parameters of multiple linear regression models. The least-squares method requires n observations of all considered independent variables x_j $j = 1, 2, \ldots, k$, i.e., x_{ij} $i = 1, 2, \ldots, n$, [38]. We will assume that the variables ε_i are uncorrelated random variables with zero mean value, and constant variance. Then, Equation (1) can be written in a modified form using the data from Table 2:

$$y_i = \beta_0 + \beta_1 x_{i1} + \beta_2 x_{i2} + \cdots + \beta_k x_{ik} + \varepsilon_i = \beta_0 + \sum_{j=1}^{k} \beta_j x_{ij} + \varepsilon_i, \text{ for } i = 1, 2, \ldots, n \tag{2}$$

Table 2. Data for multiple linear regression [38].

y	x_1	x_2	$\ldots$	x_k
y_1	x_{11}	x_{21}	$\ldots$	x_{1k}
y_2	x_{21}	x_{22}	$\ldots$	x_{2k}
$\vdots$	$\vdots$	$\vdots$		$\vdots$
y_n	x_{n1}	x_{n2}	$\ldots$	x_{nk}

We can formulate the object function L for the least-squares method of the model (2) in the form that ensures that the sum of squares of errors ε_i is minimized, i.e.,:

$$L = \sum_{i=1}^{n} \varepsilon_i^2 = \sum_{i=1}^{n} \left(y_i - \beta_0 - \sum_{j=1}^{k} \beta_j x_{ij} \right)^2 \tag{3}$$

The object function L must be minimized in respect of the parameters $\beta_0, \beta_1, \cdots \beta_k$. It can be written in the form of equations:

$$\left.\frac{\partial L}{\partial \beta_0}\right|_{\hat{\beta}_0, \hat{\beta}_1, \ldots, \hat{\beta}_k} = -2\sum_{i=1}^{n} \left(y_i - \hat{\beta}_0 - \sum_{j=1}^{k} \hat{\beta}_j x_{ij} \right) = 0 \left.\frac{\partial L}{\partial \beta_j}\right|_{\hat{\beta}_0, \hat{\beta}_1, \ldots, \hat{\beta}_k} = -2\sum_{i=1}^{n} \left(y_i - \hat{\beta}_0 - \sum_{j=1}^{k} \hat{\beta}_j x_{ij} \right) x_{ij} = 0 \; j = 1, 2, \ldots, k \tag{4}$$

The least-squares normal equations are obtained by simplifying Equations (4) into the form (5):

$$n\hat{\beta}_0 + \hat{\beta}_1 \sum_{i=1}^{n} x_{i1} + \hat{\beta}_2 \sum_{i=1}^{n} x_{i2} + \cdots + \hat{\beta}_k \sum_{i=1}^{n} x_{ik} = \sum_{i=1}^{n} y_i, \tag{5}$$

$$\begin{array}{ccccccc} \hat{\beta}_0 \sum\limits_{i=1}^{n} x_{i1} + & \hat{\beta}_1 \sum\limits_{i=1}^{n} x_{i1}^2 + & \hat{\beta}_2 \sum\limits_{i=1}^{n} x_{i1} x_{i2} + \cdots + & \hat{\beta}_k \sum\limits_{i=1}^{n} x_{i1} x_{ik} = & \sum\limits_{i=1}^{n} x_{i1} y_i \\ \vdots & \vdots & \vdots & \vdots & \vdots, \\ \hat{\beta}_0 \sum\limits_{i=1}^{n} x_{ik} + & \hat{\beta}_1 \sum\limits_{i=1}^{n} x_{ik} x_{i1} + & \hat{\beta}_2 \sum\limits_{i=1}^{n} x_{ik} x_{i2} + \cdots + & \hat{\beta}_k \sum\limits_{i=1}^{n} x_{ik}^2 = & \sum\limits_{i=1}^{n} x_{ik} y_i \end{array} \tag{6}$$

Subsequently, estimates of regression parameters $\beta_0, \beta_1, \cdots \beta_k$ are obtained by solving these equations. It is appropriate to use matrix notation to simplify the solution of equations:

$$\mathbf{y} = \mathbf{X}\boldsymbol{\beta} + \boldsymbol{\varepsilon} \tag{7}$$

where

$$y = \begin{bmatrix} y_1 \\ y_2 \\ \vdots \\ y_n \end{bmatrix}, X = \begin{bmatrix} 1 & x_{11} & x_{12} & \cdots & x_{1k} \\ 1 & x_{12} & x_{22} & \cdots & x_{2k} \\ \vdots & \vdots & \vdots & & \vdots \\ 1 & x_{n1} & x_{n2} & \cdots & x_{nk} \end{bmatrix}, \beta = \begin{bmatrix} \beta_0 \\ \beta_1 \\ \vdots \\ \beta_k \end{bmatrix} \text{ and } \varepsilon = \begin{bmatrix} \varepsilon_1 \\ \varepsilon_2 \\ \vdots \\ \varepsilon_k \end{bmatrix}, \quad (8)$$

Then, the vector of regression coefficient β estimates is calculated as

$$\hat{\beta} = \left(X^T X\right)^{-1} \left(X^T y\right) \quad (9)$$

and it can be used in notation of a multiple regression model

$$\hat{y} = X\hat{\beta} \quad (10)$$

or in the form

$$\hat{y}_i = \hat{\beta}_0 + \sum_{j=1}^{k} \hat{\beta}_j x_{ij}, i = 1, 2, \ldots, n \quad (11)$$

The difference between the actual value y_i and the corresponding modeled value $\hat{y}_i$ is called the residual.

$$SST = SSR + SSE, \quad (12)$$

Furthermore, it is necessary to verify the suitability of the proposed multiple regression model. The first recommended test is a test to verify the existence of a linear regression relationship between the dependent variable y and the selected independent variables. The null hypothesis H_0: $\beta_0 = \beta_1 = \beta_2 = \cdots = \beta_k = 0$ will be tested against the alternative hypothesis H_1: Not all the β_i, $i = 1, 2, \cdots, k$ are zero. The test will use analysis of variance, the important calculations of which are shown in Table 3. The most important part of the test procedure is the calculation of the three sums of squares in the following form:

$$\sum_{i=1}^{n} (y_i - \overline{y})^2 = \sum_{i=1}^{n} (\hat{y}_i - \overline{y})^2 + \sum_{i=1}^{n} (y_i - \hat{y}_i)^2, \quad (13)$$

Table 3. Analysis of variance for significance of regression in multiple regression [38].

Source of Variation	Sum of Squares	Degrees of Freedom	Mean Square	F
Regression	SSR	k	$MSR = SSR/k$	$F = MSR/MSE$
Error or residual	SSE	$n - (k+1)$	$MSE = SSE/(n - (k+1))$	
Total	SST	$n - 1$		

H_0: $\beta_0 = \beta_1 = \beta_2 = \cdots = \beta_k = 0$ We will not reject if the calculated value of the test statistic F is less than the critical value $F_{\alpha,k,n-k-1}$, or if the calculated P-value is greater than the selected level of significance of the test α. If we do not reject the tested null hypothesis, there is no assumed linear relationship between the independent variable y and the considered independent variables x_j $j = 1, 2, \cdots, k$. To verify the significance of individual independent variables, we can perform t-tests. The null hypothesis has a form $\beta_j = 0$ and is tested against the alternative hypothesis $\beta_j \neq 0$ for $j = 1, 2, \cdots, k$. The test statistic is calculated according to the formula:

$$t = \frac{\hat{\beta}_j}{\sqrt{\hat{\sigma}^2 C_{jj}}}, \quad (14)$$

Besides the estimated value of the coefficient $\hat{\beta}_j$, we also use the values of a standard error $\sqrt{\hat{\sigma}^2 C_{jj}}$ where C_{jj} are diagonal elements of the matrix $\left(\mathbf{X}^T\mathbf{X}\right)^{-1}$. We reject the null hypothesis if the value of the test statistic is greater than the critical value $t_{\alpha/2,n-k-1}$ or if the p-value is less than the chosen level of significance α. Because we want to evaluate the suitableness of a proposed multiple linear regression model, we use the mean square error—MSE defined as:

$$MSE = \frac{SSE}{n-(k+1)}, \tag{15}$$

The lower the MSE values, the better the regression model expresses the measured data. The same is true for the square root of MSE, which is called the standard error of estimate and is marked s. Using the multiple coefficient of determination R^2, we can calculate the share of the variability of the dependent variable y, which is expressed by the model, i.e., a combination of selected independent variables used in the regression model. It can be written in the form:

$$R^2 = \frac{SSR}{SST} = 1 - \frac{SSE}{SST}, \tag{16}$$

At best, it is equal to $R^2 = 1$ or expressed in a percentage $R^2 = 100\%$. We can use the adjusted multiple coefficient of determination R^2_{adj} to consider the number of independent variables in the proposed linear regression model. Thus, we take into account not only the values of SSE and SST, but also the numbers of freedom degrees $n-(k+1)$ and $(n-1)$ in this value calculation:

$$R^2_{adj} = 1 - \frac{\frac{SSE}{n-(k+1)}}{\frac{SST}{n-1}} \tag{17}$$

The comparison of the calculated values of R^2 and R^2_{adj} is also helpful from the point of view of considerations about the inclusion of individual independent variables in the multiple linear regression model. If their values differ significantly after the inclusion of a new variable, it is clear that the inclusion of this variable in the model is not necessary [39,40].

4. Results and Discussion

This chapter describes the proposal of multiple linear regression models for modeling temperatures in the gasification channel and in the coal, which are differed from each other in the number of independent variables considered. The coefficients of the proposed models were calculated by using the measured experimental data from the first experiment (see Section 2.1). Measured experimental data from the first experiment (i.e., the exhaust fan motor frequency, the flow of air and oxygen, and calorific value) are shown in Figure 8. Verification of the quality of the proposed models was performed on the measured data from this experiment by calculation of multiple coefficients of determination and standard error of the estimate. Furthermore, the change of experimental conditions was tested, i.e., the proposed models based on data of the first experiment were applied to the measured data from the second experiment (see Section 2.2), which were obtained under different conditions. Measured experimental data from the second experiment (i.e., the exhaust fan motor frequency, the flow of air and oxygen, and calorific value) are shown in Figure 9.

4.1. The Proposal of a Multiple Linear Regression Model for the Channel's and Coal's Temperatures

In the first stage of the solution, the kind of independent variables was chosen for the proposal of the temperature prediction model. These variables represent the measured variables: frequency of the fan located behind the ex situ reactor (see Figure 2), airflow, oxygen flow, syngas calorific value, temperatures measured in the channel and in coal. The models for the individual dependent variables—channel temperatures (i.e., T_j^{mod} $j = 3, 4, \cdots, 13$) differed by the selected measured temperatures in order from the place the coal seam ignition (i.e., gasification agent input) to the calculated temperature. For the model of the dependent variable temperature T_3^{mod}, only the temperatures T_1, T_2, T_{14}, and T_{15} were taken into account, but, e.g., for the temperature T_{12}^{mod}, all previously measured channel temperatures T_3 to T_{11} were included in the model. The results of the correlation analysis (i.e., mainly calculations R^2 and R^2_{adj} coefficients) showed that temperatures measured in coal (i.e., temperatures T_{16}, T_{17}, T_{18}, T_{19}, T_{20}, T_{21}) are insignificant independent variables for the model proposal, and due to this reason, were excluded from the models. The mathematical form of the proposed regression models for channel's temperatures can be written in general form, as follows:

$$T_j^{mod} = b_0 + \sum_{i=1}^{7} b_i \cdot x_i + \sum_{i=8}^{i+j-3} b_i \cdot x_i, \tag{18}$$

where: T_j^{mod} are modelled temperatures for $j = 3, 4, \cdots, 13$ (°C); x_1 is frequency (Hz); x_2 is airflow (m^3.h^{-1}); x_3 is oxygen (m^3.h^{-1}); x_4 is calorific value (MJ.m^{-3}); x_5 is temperature T_1 (°C); x_6 is temperature T_{14} (°C); x_7 is temperature T_{15} (°C); x_i is temperature T_{i-6} for $i = 8, 9, \cdots, 18$ (°C).

Measured data from the first experiment were used to determine the coefficients of multiple linear regression models. The calculations were performed in the Minitab statistical software using the least-squares method. The calculated coefficients of regression models for channel temperatures are shown in Table 4.

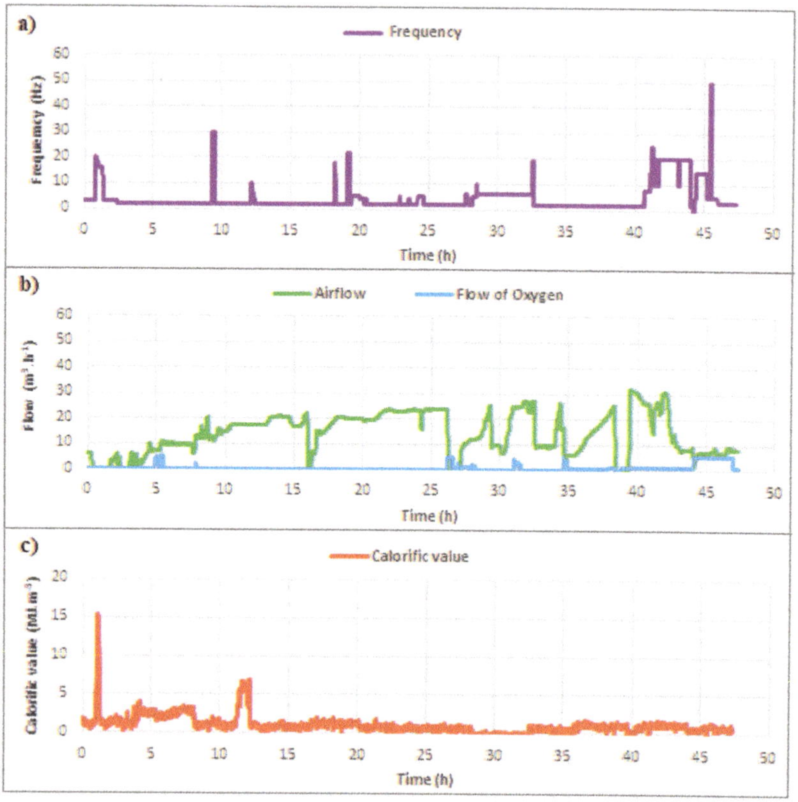

Figure 8. The measured data behaviors in the first experiment: (**a**) the exhaust fan motor frequency, (**b**) the air and oxygen, and (**c**) the calorific value (Source: own elaboration).

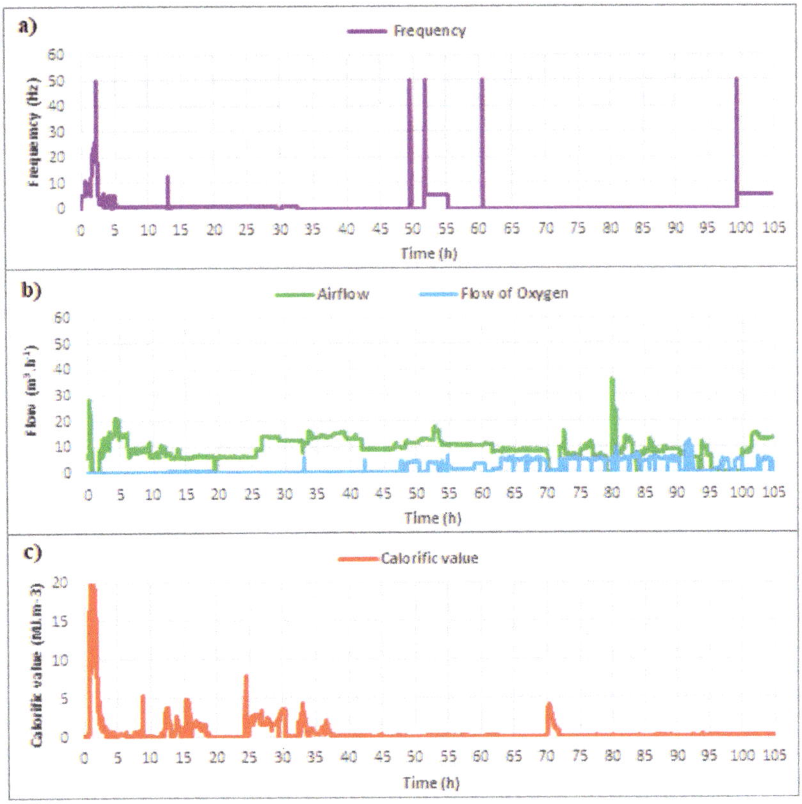

Figure 9. The measured data behaviors in the second experiment: (**a**) the exhaust fan motor frequency, (**b**) the air and oxygen, and (**c**) the calorific value (Source: own elaboration).

Table 4. Coefficients of regression models for channel temperatures and values of a multiple coefficient of determination and a standard error of the estimate (Source: own elaboration).

Coefficients	T_3^{mod}	T_4^{mod}	T_5^{mod}	T_6^{mod}	T_7^{mod}	T_8^{mod}	T_9^{mod}	T_{10}^{mod}	T_{11}^{mod}	T_{12}^{mod}	T_{13}^{mod}
b_0	122.473	67.662	38.014	15.472	4.003	96.044	−24.393	88.273	71.571	−17.614	−28.813
b_1	−0.824	−0.287	0.352	0.519	0.314	−0.250	−0.002	1.218	0.038	−1.027	−0.109
b_2	−1.821	1.339	1.141	−0.684	0.127	1.239	0.068	2.759	0.077	2.409	0.193
b_3	8.895	8.739	2.781	−0.569	−1.390	2.237	0.042	4.450	−0.298	−2.911	−0.255
b_4	1.617	−4.234	−0.241	−2.720	−0.051	4.504	0.143	0.369	−3.353	2.398	0.092
b_5	−0.419	0.173	0.052	0.063	−0.056	−0.208	−0.011	−0.726	0.155	0.170	0.015
b_6	0.210	−0.106	0.035	−0.052	0.117	0.038	0.001	0.403	−0.227	−0.187	−0.020
b_7	0.174	−0.005	−0.099	−0.042	−0.022	−0.003	−0.002	0.128	0.026	−0.001	0.000
b_8	0.936	−0.082	−0.055	0.279	−0.040	0.070	0.001	−0.146	0.197	0.030	0.006
b_9		0.971	0.186	−0.140	−0.110	0.083	0.010	0.609	−0.365	−0.247	−0.023
b_{10}			0.815	−0.315	0.211	−0.081	−0.002	1.067	0.096	0.053	0.002
b_{11}				1.187	−0.428	0.079	−0.004	−0.763	−0.205	−0.028	0.012
b_{12}					1.346	−0.092	−0.006	−0.553	0.494	−0.086	−0.023
b_{13}						0.931	−0.944	−0.004	0.074	0.294	0.031
b_{14}							1.944	0.644	−0.199	−0.135	−0.013
b_{15}								0.303	0.441	−0.188	−0.012
b_{16}									0.351	0.267	0.023
b_{17}										0.981	−0.921
b_{18}											1.912
s	44.053	28.864	19.316	14.350	11.790	19.647	6.139	30.922	18.581	14.395	6.182
R^2	96.11%	98.04%	99.07%	99.50%	99.67%	98.91%	99.87%	96.48%	98.60%	98.23%	99.86%
R^2_{adj}	96.10%	98.03%	99.06%	99.49%	99.67%	98.90%	99.87%	96.46%	98.59%	98.21%	99.86%

Tests of the null hypothesis $\beta_0 = \beta_1 = \beta_2 = \cdots = \beta_k = 0$ at the selected level of significance 0.05 realized by Analysis of Variance for Significance of Regression in Multiple Regression, we concluded for all proposed regression models by rejecting this hypothesis and not rejecting the alternative hypothesis, H_1: Not all the β_i, $i = 1, 2, \cdots, k$ are zero, and thus confirming the existence of a multiple linear relationship between the independent variable and the dependent variables.

Table 4 shows that the influence of individual independent variables included in the regression models for calculating channel temperatures is not unambiguous but varies in terms of force (i.e., size of the coefficient) and terms of type (i.e., direct/indirect dependence). The last three lines of Table 4 contain selected results of correlation analysis, namely the standard error of estimate—s, the multiple coefficient of determination—R^2 and the adjusted multiple coefficient of determination—R^2_{adj}.

The standard error of estimate—s for individual models ranges from 6.1390 to 44.0534, which is an acceptable result due to the size of the measured temperatures used (i.e., maximal is circa 1200 ° C). Values of R^2 and R^2_{adj}, ranging from 96.11% (i.e., for T_3) to 99.87% (i.e., for T_9), clearly show that each of the proposed regression models represents more than 96% of the variability of the dependent variable (i.e., channel temperature). Thus, it can be stated that the proposed regression models are suitable for use. The behavior of measured T_j and modelled T_j^{mod} temperature (i.e., for temperatures T_3 and T_9) is shown in Figure 10. The difference between the calculated values R^2 and R^2_{adj} is minimal due to the large number of data (i.e., $n = 2846$), and therefore only the value R^2_{adj} is shown in the following tables.

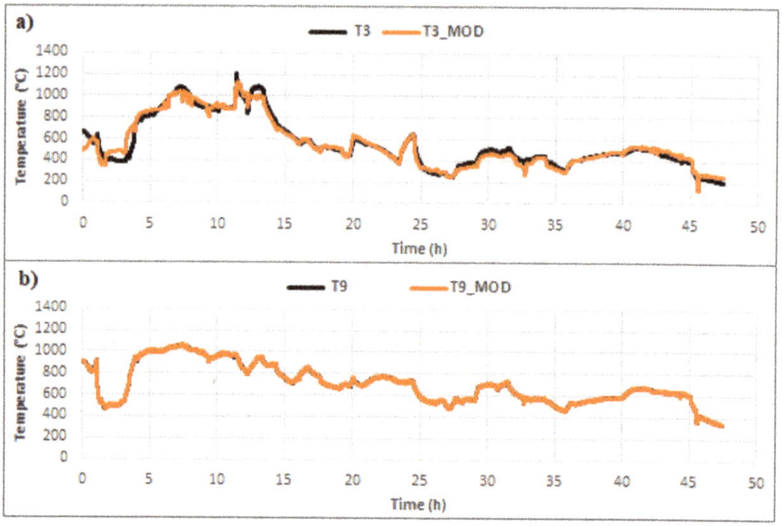

Figure 10. The measured (T) and modelled (T_MOD) temperature behavior, (**a**) temperature T_3, (**b**) temperature T_9 (Source: own elaboration).

In the next phase, we focused on solving temperatures in coal, i.e., temperatures T_{17} and T_{18}. The proposed regression model included channel temperatures T_1 to T_7, i.e., temperatures located from the ignition place of the coal seam to the cross-section of the generator where the temperatures T_{17} and T_{18} were measured. The results of the correlation analysis (i.e., values R^2 and R^2_{adj}) showed that temperatures measured in coal (i.e., temperatures T_{16} and T_{19}) located in the same section as modeled temperatures are insignificant for the proposal of the model, and due to this reason were excluded from the models. The mathematical form of the proposed regression models for temperatures T_{17} and T_{18} can be written in general form, as follows:

$$T_j^{mod} = b_0 + \sum_{i=1}^{13} b_i \cdot x_i \tag{19}$$

where: T_j^{mod} are modelled temperatures for $j = 17,\ 18$ (°C); x_1 is frequency (Hz); x_2 is airflow (m^3.h^{-1}); x_3 is oxygen (m^3.h^{-1}); x_4 is calorific value (MJ.m^{-3}); x_5 is temperature T_1 (°C); x_6 is temperature T_{14} (°C); x_7 is temperature T_{15} (°C); x_i is temperature T_{i-6} for $i = 8, 9, \cdots, 13$ (°C).

The calculated coefficients for individual models are shown in Table 5. The table shows that the temperature T_7 has the most significant influence on the modeled temperatures T_{17}^{mod} (see Figure 11a) and T_{18}^{mod} (see Figure 11b) of all considered measured temperatures (i.e., independent variables). The calorific value and airflow have a more significant effect on the modelled temperature T_{18}^{mod}, which can be caused by the non-uniform gasifying coal along the right and left sides of the gasification channel.

Table 5. Coefficients of regression models for coal temperatures and values of multiple coefficient of determination and standard error of the estimate (Source: own elaboration).

Coefficients	T_{17}^{mod}	T_{18}^{mod}
b_0	233.232	373.710
b_1	1.119	−0.343
b_2	−1.952	−9.548
b_3	5.449	1.597
b_4	−7.515	−20.216
b_5	−0.957	−0.245
b_6	0.467	−0.283
b_7	0.098	−0.038
b_8	0.035	0.301
b_9	0.649	−0.091
b_{10}	0.750	−0.345
b_{11}	0.155	3.187
b_{12}	0.539	1.082
b_{13}	−1.286	−3.587
s	66.82	115.56
R^2	86.00%	62.17%
R^2_{adj}	85.94%	61.98%

Table 5 shows that the values of the standard error of estimate—s are higher (i.e., 66.82 and 115.56), and at the same time, the values R^2 (i.e., 86.00 and 62.17%) and R^2_{adj} (i.e., 62.17 and 61.98%) are lower than at the modeled channel temperatures. This result indicates a worse prediction of coal temperatures by proposed regression models. A significant difference can be seen between correlation characteristics for temperature T_{17} and T_{18}, which also supports the previous conclusion about the non-uniform gasifying coal along the sides of the gasification channel, i.e., the gasification process was faster on the temperature T_{17} side. The difference in the velocity of gasifying coal was circa 6 h. The temperature of 600 °C was reached on the left side of the gasification channel approximately at the 5th hour, while on the right side at the 11th hour. The difference of the burning velocities can be caused by uniform leakage of gasification agent through the upper edge of the ex situ reactor or by the fall of the overburden layers into the space of the formed cavity.

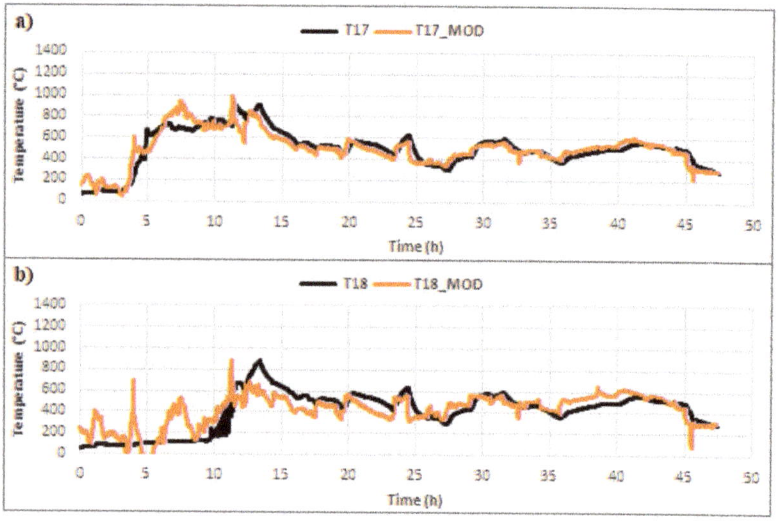

Figure 11. The measured (T) and modelled (T_MOD) temperature behavior, (**a**) temperature T_{17}, (**b**) temperature T_{18} (Source: own elaboration).

4.2. *Coefficients Application of Multiple Linear Regression Models on Data from the Second Experiment*

Multiple linear regression models proposed for the data from the first experiment were verified on the data from the second experiment. This experiment differed in the amount of gasified coal (see Section 2.2) and the gasification time. Proposed regression models were applied only to the measured data from the second experiment in the first phase of verification. Modeled temperature values from the proposed regression models were used to calculate temperatures depending on their values during the second verification phase. For example, the modeled temperature T_3 (i.e., T_3^{mod}) was used in the temperature calculation T_4, and then the modeled temperature values T_3 (i.e., T_{13}^{mod}) and T_4 (i.e., T_4^{mod}) were used to calculate the temperature T_5, etc. The goal of this phase was to verify whether it is possible to use only measured temperatures at the input to the ex situ reactor (i.e., T_1, T_2, T_{14}, and T_{15}) to calculate temperatures in the gasified coal seam model (i.e., from T_3 to T_{13}, and T_{17} and T_{18}). It was assumed that in a real gasified coal seam are not suitable conditions for measuring temperatures along the length of this coal seam.

The graph shown in Figure 12 contains selected results of correlation analysis for both phases, i.e., R^2_{adj1} values for the first phase and R^2_{adj2} values for the second phase. Assuming that in the best case, the modeled values exactly match the observed values, $R^2 = 1$ or 100%, in the opposite case, i.e., if $R^2 = 0$, modeled values do not correspond to the measured values at all. The deviation between the measured and modeled values in some cases was so large that R^2_{adj} values were not within the specified range because coefficients calculated from the measured values of the 1st experiment were applied to the values of the 2nd experiment. Therefore, these values were replaced by 0% in the graph.

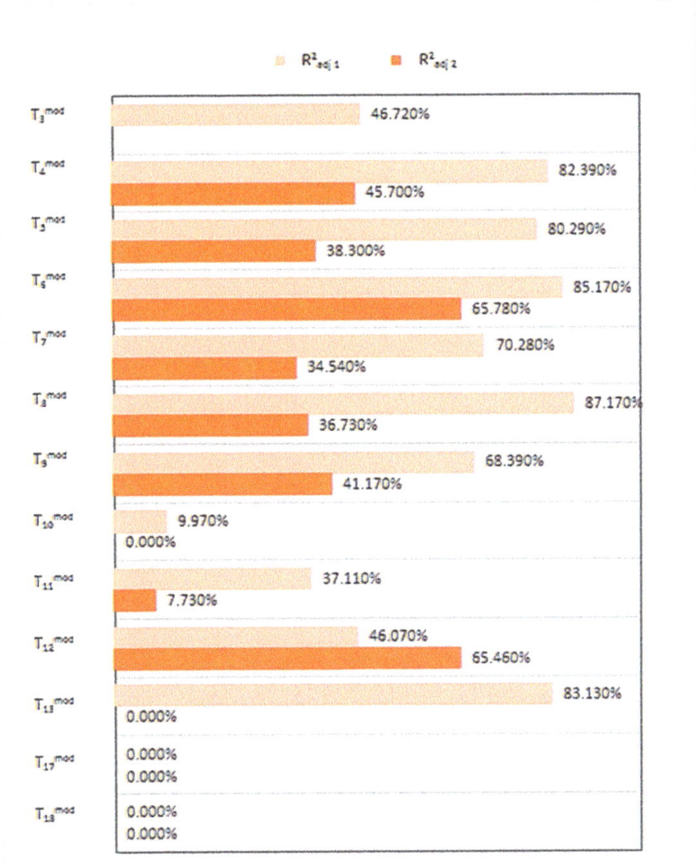

Figure 12. Values R^2_{adj} of proposed regression models for the first phase—R^2_{adj1} and the second phase—R^2_{adj2} (Source: own elaboration).

The values of the multiple coefficient of determination for channel temperatures at the first phase are significantly lower than their values in the first experiment and differ significantly from each other. The maximum value of the multiple coefficient of determination was reached at the measured temperature T_8 (see Figure 13a). Thus, it was possible to represent 87.17% of the variance of the original variable by its regression model. The smallest value of the multiple coefficient of determination was reached for the measured temperature T_{10} (see Figure 13b) when it was possible to represent only 9.97% of the variance of the original variable.

It can be observed in Figure 14a,b differences in the progress of the burning coal on the right and left sides of the ex situ. At the same time, it is possible to see that twice the volume of coal in the ex situ reactor caused the application of the proposed models from the first experiment for coal temperatures T_{17} and T_{18} shows more significant deviations than for channel temperatures. The regression model of temperature T_{17} in the second half-time of the experiment shows significantly lower deviations to the measured values than in its first half-time (see Figure 14a).

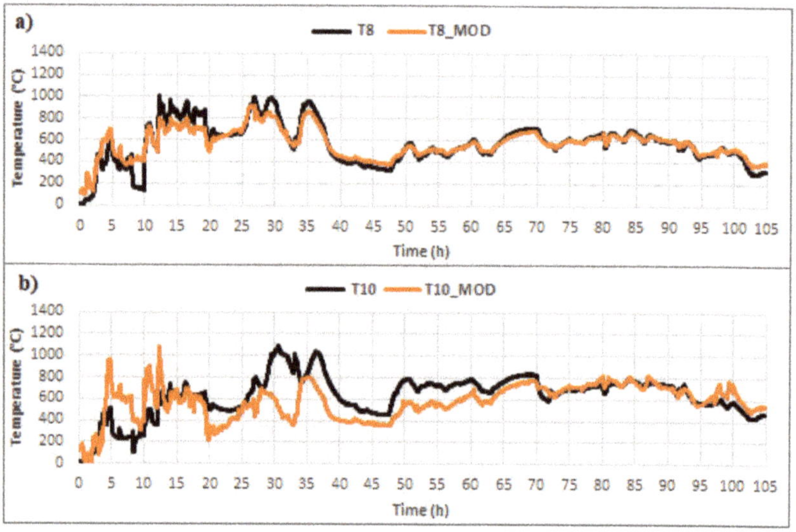

Figure 13. The measured (T) and modeled (T_MOD) temperature behavior, (**a**) temperature T_8, (**b**) temperature T_{10} (Source: own elaboration).

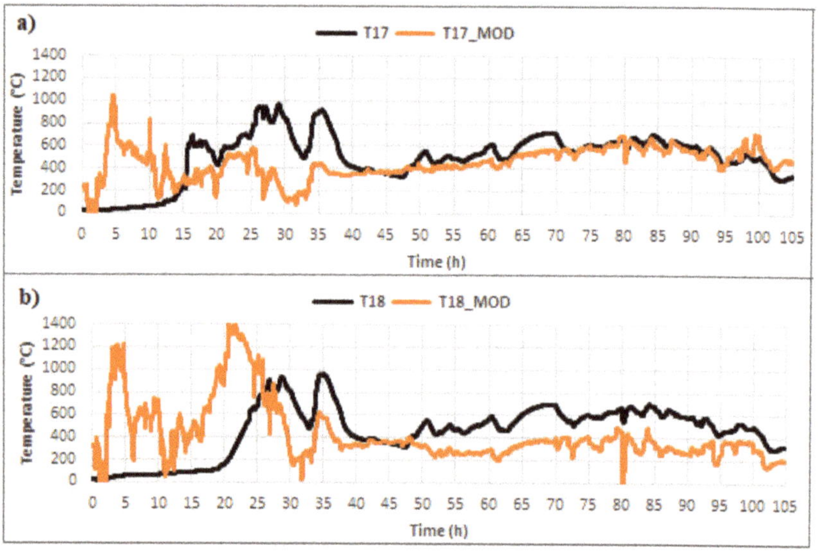

Figure 14. The measured (T) and modeled (T_MOD) temperature behavior in the first phase verification, (**a**) temperature T_{17}, (**b**) temperature T_{18} (Source: own elaboration).

The application of regression models using the modeled temperatures in the second phase verification showed that multiple coefficients of determination are even lower than for the first phase. The maximum value of 65.78% was obtained for temperature T_6 (see Figure 15a), and the minimum value of 7.73% was obtained for temperature T_{11} (see Figure 15b). The zero values R^2_{adj} for temperatures T_{10}, T_{13}, T_{17}, and T_{18} indicate a low agreement of measured and model values, i.e., the unsuitability of using the proposed regression models obtained from the first experiment for modeling temperatures in the second experiment. The graphs of temperatures T_{17} (see Figure 16a) and for T_{18} (see Figure 16b) are shown to a better visual representation.

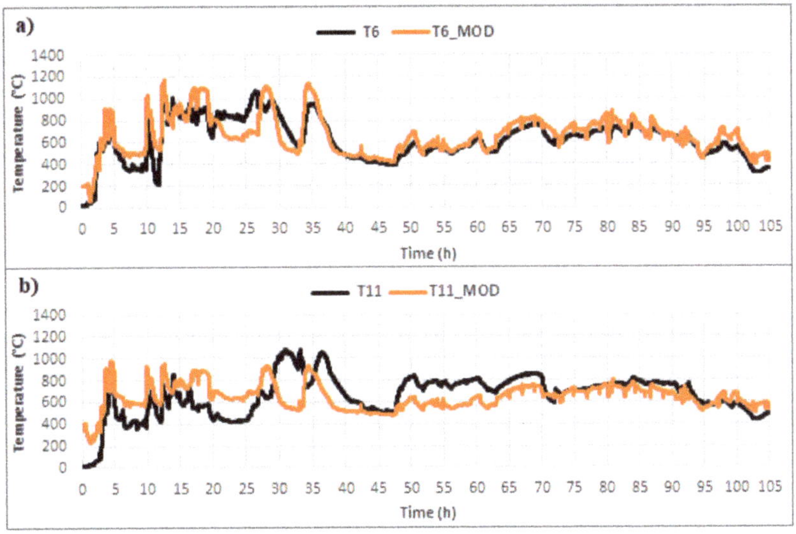

Figure 15. The measured (T) and modeled (T_MOD) temperature behavior, (**a**) temperature T_{16}, (**b**) temperature T_{11} (Source: own elaboration).

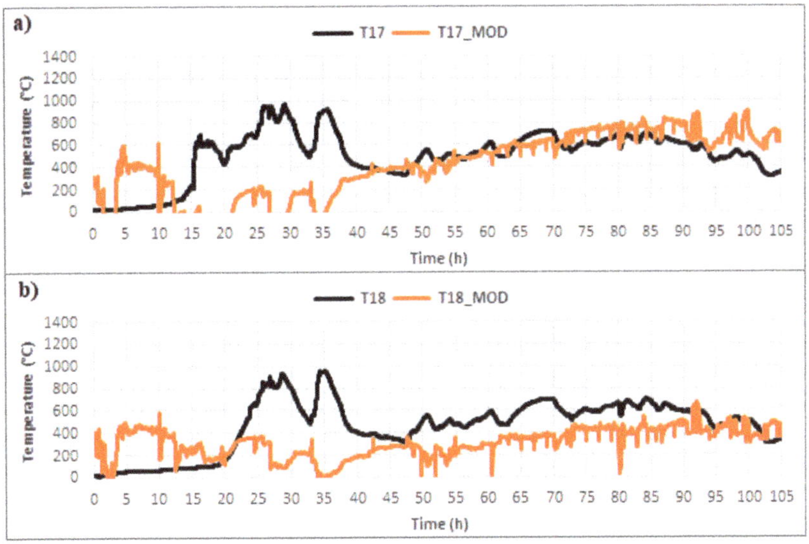

Figure 16. The measured (T) and modeled (T_MOD) temperature behavior in the second phase verification, (**a**) temperature T_{17}, (**b**) temperature T_{18} (Source: own elaboration).

The conditions of the second experiment realization were different from the first experiment's conditions due to the volume of gasified coal and the insulating materials used. Therefore, it was difficult to apply the proposed regression models based on data from the first experiment for data from the second experiment. The use of modeled temperatures in the second phase verification instead of those measured caused the deviations transfer of dependent variable values to the calculations of other temperatures. For this reason, we focused on creating regression models with a lesser number of independent variables in the next solution.

4.3. The Proposal of a Modified Multiple Linear Regression Model for the Channel's and Coal's Temperatures

We tested several multiple linear regression models calculated based on the measured values from the first experiment to optimize the number of independent variables. These tests have differed from each other in the number of independent variables involved in regression models. The most suitable type of regression model for modeling channel temperatures T_3 to T_{13} in terms of the minimum value of the multiple coefficient of determination proved to be a model including only two independent variables, namely calorific value and temperature (e.g., T_2 for T_3 calculation, T_3 for T_4 calculation, etc.). The proposed type of regression model can be written in the following form:

$$T_j^{mod} = b_0 + b_1 \cdot x_1 + b_2 \cdot x_2 \tag{20}$$

where: T_j^{mod} are modeled temperatures for $j = 3, 4, \cdots, 13$ (°C); x_1 is calorific value (MJ.m^{-3}); x_2 is temperature T_{j-1} (°C).

The calculated coefficients for individual regression models of modelled channel temperatures and selected correlation characteristics (i.e., s and R^2_{adj}) are shown in Table 6. The effect of calorific value expressed by the individual calculated coefficients b_1 is predominantly indirect because the calculated coefficients have mostly negative signs. The effect of temperature is always direct, which can be seen in the positive values of the calculated coefficients b_2.

Table 6. Coefficients of modified regression models for channel temperatures and values of a multiple coefficient of determination and a standard error of the estimate (Source: own elaboration).

Predicted Temperature	b_0	b_1	b_2	s	R^2_{adj}
T_3^{mod}	106.564	−2.890	0.855	64.625	91.60%
T_4^{mod}	107.575	−4.334	0.918	34.127	97.20%
T_5^{mod}	46.542	0.153	0.962	24.200	98.50%
T_6^{mod}	1.378	1.952	0.998	28.016	98.10%
T_7^{mod}	−3.909	0.799	1.011	16.123	99.40%
T_8^{mod}	105.407	−3.932	0.912	27.346	97.90%
T_9^{mod}	70.103	−5.691	0.928	29.542	97.10%
T_{10}^{mod}	39.156	−2.646	0.944	17.296	98.90%
T_{11}^{mod}	31.217	−3.239	0.953	18.223	98.60%
T_{12}^{mod}	4.998	−5.582	0.993	33.147	95.60%
T_{13}^{mod}	−25.378	−6.724	1.039	32.607	96.10%

Tests of the null hypothesis $\beta_0 = \beta_1 = \beta_2 = \cdots = \beta_k = 0$ at the selected level of significance 0.05 realized by Analysis of Variance for Significance of Regression in Multiple Regression, we concluded for all proposed regression models by rejecting this hypothesis and not rejecting the alternative hypothesis, H_1: Not all the β_i, $i = 1, 2, \cdots, k$ are zero, and thus confirming the existence of a multiple linear relationship between the independent variable and the dependent variables.

The standard error of estimate—s for individual models ranges from 17.296 to 34.127, only for temperature T_3 is higher (i.e., 64.625). It is an acceptable result due to the size of the measured temperatures (i.e., maximal is circa 1200 °C). Values of R^2_{adj}, ranging from 91.60% (i.e., for temperature T_3) to 99.40% (i.e., for temperature T_7), clearly show that each of the proposed regression models represents more than 91% of the variability of the dependent variable (i.e., channel temperature). Thus, it can be stated that the proposed regression models are suitable for use. The behavior of measured T_j and modeled T_j^{mod} temperature (i.e., for temperatures T_3 and T_7) is shown in Figure 17a,b.

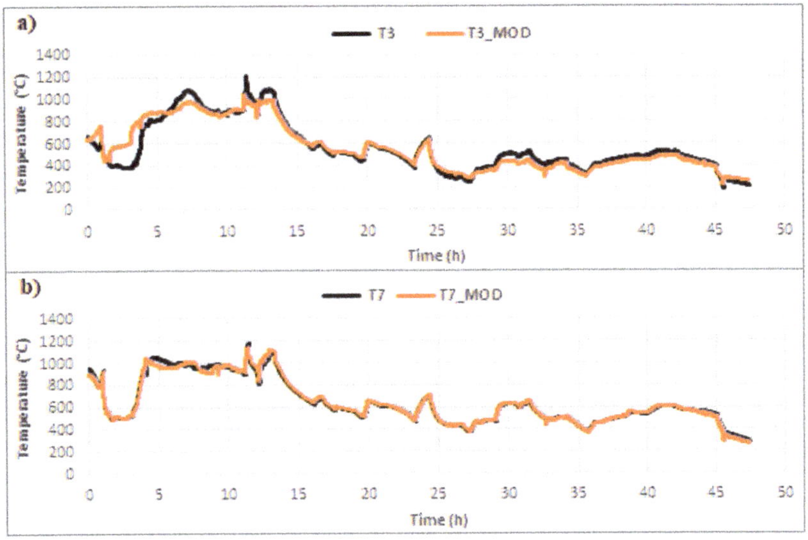

Figure 17. The measured (T) and modeled (T_MOD) temperature behavior, (**a**) temperature T_3, (**b**) temperature T_7 (Source: own elaboration).

After analyzing several variants of the solution, a model for temperatures in coal (i.e., for temperatures T_{17} and T_{18}) was proposed. This model includes four independent variables, i.e., calorific value, channel temperature T_7 and temperatures T_{16} and T_{19} measured in the insulation layer at the edges of the ex situ reactor. The solution was based on the assumption of measuring temperatures in the surrounding rocks of the gasified coal seam. The proposed model can be written in the following form:

$$T_j^{mod} = b_0 + b_1 \cdot x_1 + b_2 \cdot x_2 + b_3 \cdot x_3 + b_4 \cdot x_4, \tag{21}$$

where: T_j^{mod} are modelled temperatures for $j = 17,\ 18$ (°C); x_1 is calorific value (MJ.m^{-3}); x_2 is temperature T_7 (°C); x_3 is temperature T_{16}; x_4 is temperature T_{19} (°C).

The calculated coefficients for individual regression models of modeled coal temperatures and selected correlation characteristics (i.e., s and R^2_{adj}) are shown in Table 7. The modified regression model of temperature T_{17} (see Figure 18a) worse represents measured temperature T_{17} in the first 10 h of the experiment while compared to the regression model results shown in Figure 11a. This conclusion is also confirmed by the reduction of R^2_{adj} values from 85.94 to 71.48%. The modified regression model of temperature T_{18} (see Figure 18b) better represents measured temperature T_{18} during the experiment than the regression model shown in Figure 11b. Proof that this is an increase in R^2_{adj} values from 61.98 to 95.65%.

Table 7. Coefficients of modified regression models for coal temperatures and values of a multiple coefficient of determination and a standard error of the estimate (Source: own elaboration).

Predicted Temperature	b_0	b_1	b_2	b_3	b_4	s	R^2_{adj}
T_{17}^{mod}	−87.233	0.110	0.543	0.414	0.156	95.179	71.48%
T_{18}^{mod}	9.627	12.987	−0.031	0.466	0.533	39.866	95.65%

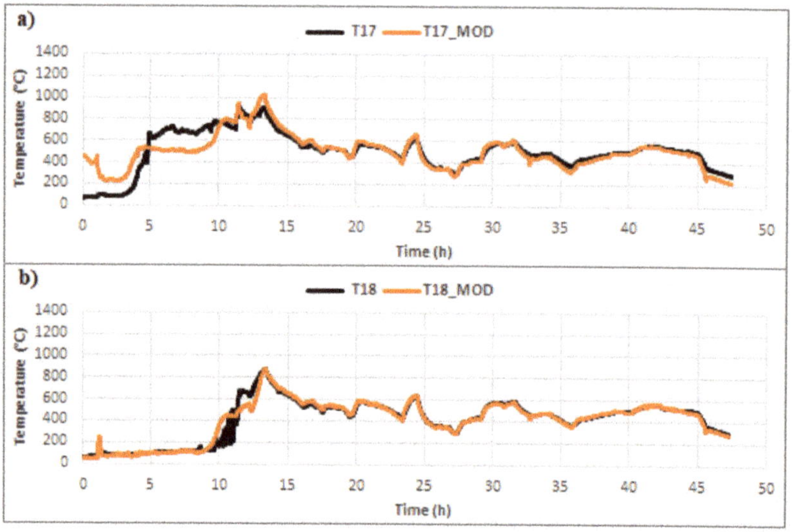

Figure 18. The measured (T) and modeled (T_MOD) temperature behavior by using modified regression models, (**a**) temperature T_{17}, (**b**) temperature T_{18} (Source: own elaboration).

4.4. Coefficients Application of Modified Multiple Linear Regression Models on Data from the Second Experiment

The modified multiple linear regression models proposed for the data from the first experiment were verified on the data from the second experiment similarly as in the case of the verification of regression models described in Section 4.2. Two phases were used in the verification, similar to the previous cases. At first, the calculations were performed only with the measured data and subsequently also with calculated. Selected results of the correlation analysis—R^2_{adj} values for all modeled temperatures are shown in Figure 19.

The first verification phase's multiple coefficient of determination values R^2_{adj1} were in a range from 77.15% (i.e., temperature T_{13}) to 94.99% (i.e., temperature T_3) for channel temperatures calculated only from the measured temperatures. It is possible to observe a decreasing trend of these values towards the ex situ reactor output based on these values. The behavior of measured and modelled temperatures T_3 and T_{13} is shown in Figure 20a,b.

It is visible a significant improvement in the representation of measured coal temperatures by modeled coal temperatures in compared temperature behaviors in Figure 21a,b with temperature behaviors in Figure 14a,b. This result is also confirmed by the achieved values R^2_{adj1}, i.e., 48.936% for temperature T_{17} and 48.471% for temperature T_{18} (see Figure 19).

The second verification phase's multiple coefficient of determination values R^2_{adj2} were in a range from 8.54% (i.e., temperature T_9) to 83.08% (i.e., temperature T_4) for channel temperatures calculated from the modeled temperatures. It is possible to observe a decrease in these values towards the ex situ reactor output based on these values. This decrease is due to reducing the number of independent variables, i.e., by minimization of the transmitted calculation error. We can state that the applicability of the proposed models is sufficient for temperatures T_4 to T_6 by observing the values of R^2_{adj2} for temperatures measured in the gasification channel. The behavior of measured and modeled temperatures T_4 and T_9 is shown in Figure 22a,b.

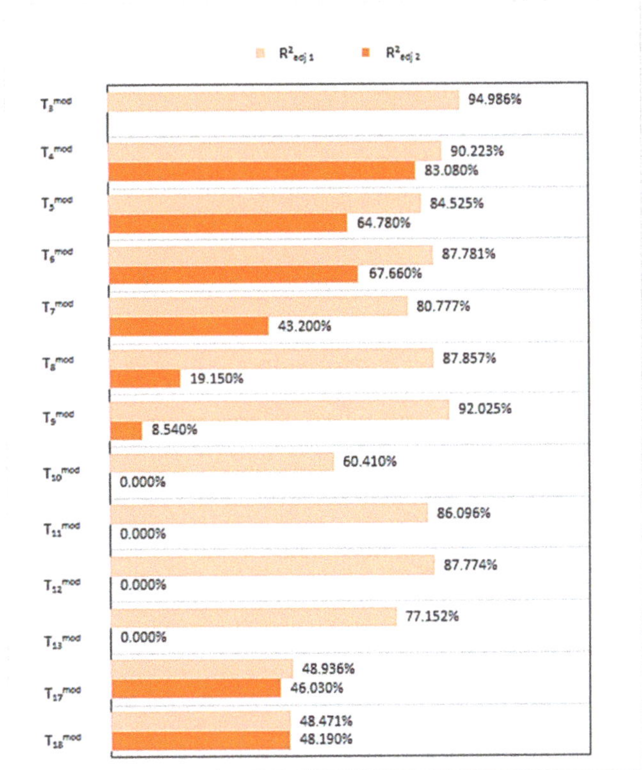

Figure 19. Values R^2_{adj} of modified regression models for the first phase—R^2_{adj1} and the second phase—R^2_{adj2} (Source: own elaboration).

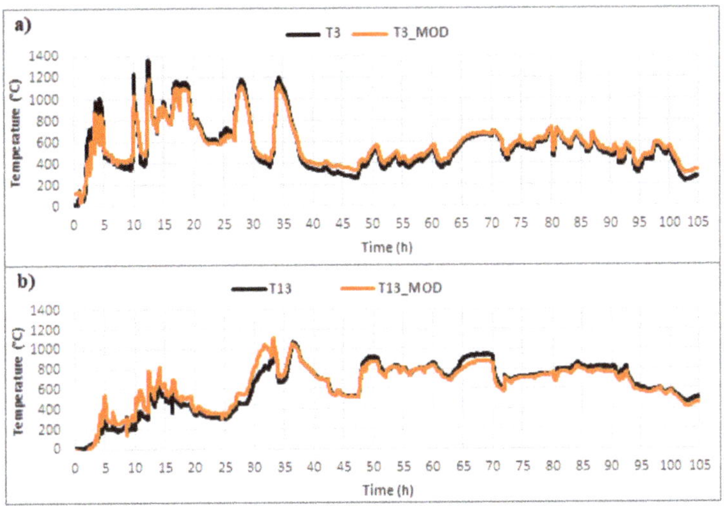

Figure 20. The measured (T) and modeled (T_MOD) temperature behavior, (**a**) temperature T_3, (**b**) temperature T_{13} (Source: own elaboration).

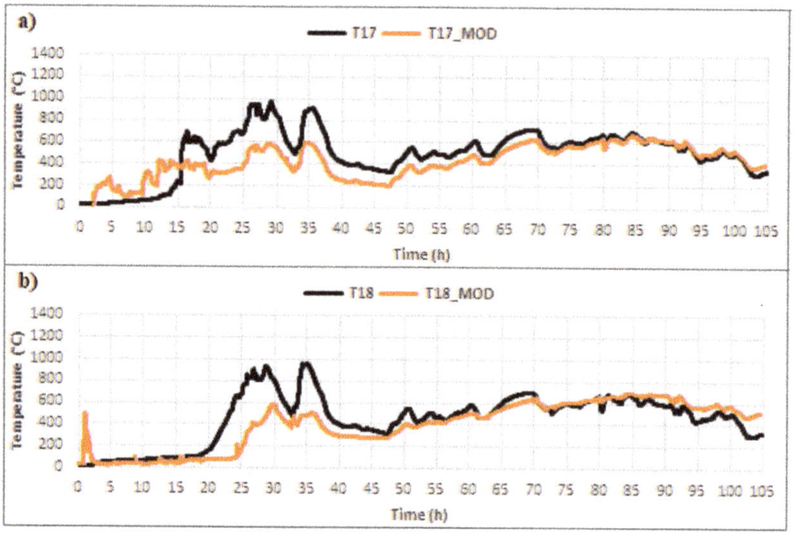

Figure 21. The measured (T) and modeled (T_MOD) temperature behavior in the first phase verification by using modified regression models, (**a**) temperature T_{17}, (**b**) temperature T_{18} (Source: own elaboration).

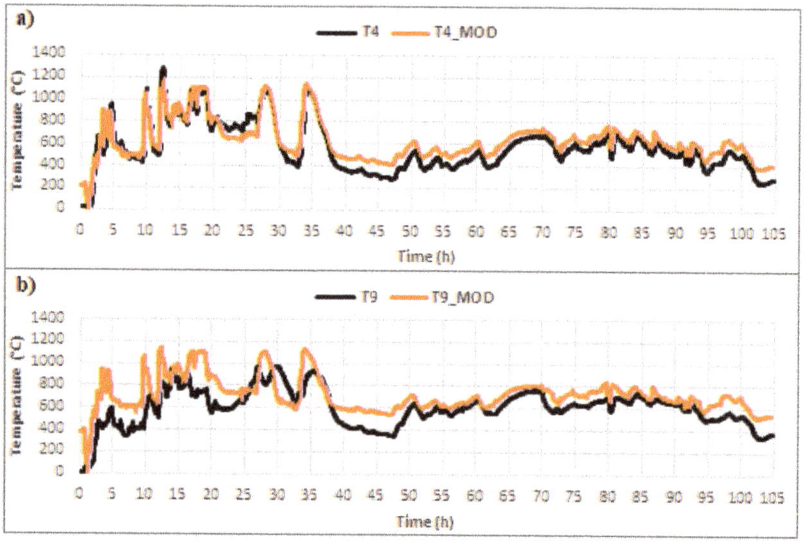

Figure 22. The measured (T) and modeled (T_MOD) temperature behavior, (**a**) temperature T_4, (**b**) temperature T_9 (Source: own elaboration).

Results showed a significant improvement in the representation of measured values by modeled values at comparison modeled coal temperatures showed in Figure 23a,b to modeled coal temperatures showed in Figure 16a,b. It is confirmed by the achieved values $R^2_{adj2} = 46.03\%$ for temperature T_{17} and 48.19% for temperature T_{18}. The achieved values of R^2_{adj} in the first and second phases have differed only minimally, i.e., the replacement measured values by calculated values did not reduce the expression quality of the dependent variable variance (i.e., temperature T_{17} or T_{18}). Reach values R^2_{adj}, around 50%, are low in terms of the suitability of these models. We can say that the deviations between the measured and modelled values of the dependent variables T_{17} and T_{18} are significantly lower from the 45th hour of the experiment. Deviations of these temperatures in the first half of the experiment cause small values of R^2_{adj2} (see Figure 23a,b). The similarity of temperature behaviors between Figures 21b and 23b is caused by lower value regression coefficient b_2 (i.e., −0.031). This coefficient lower value reduces the effect of the temperature T_7 (i.e., measured in the first phase verification and modelled in the second phase verification, at the temperature T_{18} calculation) on the temperature T_{18}.

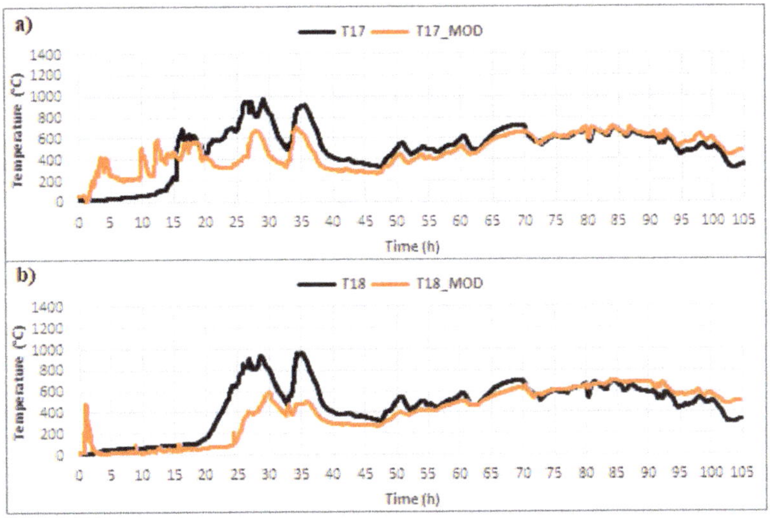

Figure 23. The measured (T) and modeled (T_MOD) temperature behavior in the second phase verification by using modified regression models, (**a**) temperature T_{17}, (**b**) temperature T_{18} (Source: own elaboration).

The average calorific value for the first experiment was 1.155 and for the second experiment was 0.657. Thus, we can say that the influence of calorific value on the modeled coal temperatures T_{17} and T_{18} was reduced by almost half in the second experiment. The calorific value reduction causes significant differences between the measured and modeled temperatures, especially if the temperatures are above 600 °C. Syngas with the desired composition and calorific value can be produced (i.e., the transformation of coal into gas occurs) due to this temperature.

5. Conclusions

This described research aimed to propose regression models for modeling temperatures in the gasification channels and the coal seam gasified in the ex situ reactor. The proposed models were to contribute to developing a methodology for predicting temperatures in a gasified coal seam. Improving the prediction of these temperatures with higher accuracy makes it possible to identify places in the coal seam where coal to gas is transformed and the underground cavity is formed. The prediction of coal seam temperatures would also allow the development of methods to control the UCG process based on modeled temperatures in the coal seam.

Two experiments were performed for the proposal and verification of regression models. These experiments were differed by the volume of gasified coal and thus also in the duration of the experiment. The proposal of regression coefficients was performed on the data from the first experiment, but the verification of the proposed regression models was performed mainly on the data from the second experiment. The ability to use the created regression models to predict temperatures in the UCG process realized under approximately the same geological conditions, e.g., the same structure of the coal seam with the surrounding rocks, was tested. The coal model (i.e., placement of coal blocks and isolation layers) in the second experiment had the same structure as the coal model of the first experiment, where the difference was the amount of gasified coal. This coal amount affected the experiment duration, the amount of gasification agent used, and the output data obtained from the experiment.

The quality of the models was assessed by calculating the multiple coefficient of determination and the standard error of the estimate. In the first stage of the research, two model structures were proposed, i.e., multiple linear regression models for channel and coal temperatures. A more number of independent variables influencing the gasification process were considered in the first solution. The multiple coefficient of determination of the proposed regression models for channel temperatures expressed more than 96% and for coal temperatures more than 61% of the variability of the dependent variable. This solution proved to be less efficient in verifying proposed models on the data from the second experiment because it was influenced by the transmission of the error from all independent variables. The multiple coefficient of determination was ranged from 9.97 to 87.17% for channel temperatures and for coal temperatures was outside its specified range when was verified on directly measured data. Verification using also modeled temperatures showed a very low similarity between the measured and modeled temperatures.

The structure of the independent variables was optimized, and subsequently, final regression models were created under their significantly smaller number. The multiple coefficient of determination of the proposed regression models for channel temperatures expressed more than 91% and for coal temperatures more than 71% of the variability of the dependent variable. Verification of the data from the second experiment confirmed the correctness of reducing the number of independently variables by increasing its values. The stability of the coal temperatures modeling was not affected by the transition from measured to modeled data at their calculation because the value of the multiple coefficient of determination decreased only minimally (i.e., for temperature T_{17} from 48.94 to 46.03% and temperature T_{18} from 48.47 to 48.19%). The results indicated the possibility of using the proposed model of channel temperatures for the first half of the generator. The prediction of coal temperatures showed a 50% similarity of measured and modeled values, i.e., use the model only on data in the second half time of the experiment. This result was caused by lower values of the measured calorific value of syngas, mainly at the internal temperature T_{18}, when the regression model contained a higher value of the regression coefficient b_1. The calorific value could be influenced by the suction air at the outlet of the ex situ reactor, where a fan was placed on improving the control of the UCG process.

Low similarities of measured and modeled temperatures and thus the low quality of the proposed regression models could be caused by leaks of gasification agent through ex situ reactor cracks. We can say that there are still many options for the development of regression models for temperature prediction in the gasified coal seam, for example:

- improving the experimental process for data collection, e.g., by reducing to a minimum respectively by removing gasification agent leaks during the experiment, continuous measurement of gas composition and calorific value along the length of the ex situ reactor, etc.
- including dimensionless numbers in the regression models proposal for their application under various conditions, e.g., the Fourier number as a dimensionless time
- determination of relevant independent variables for modeling temperatures in specific places of the coal seam by extended regression analysis

Author Contributions: Conceptualization, M.D. and M.B.; methodology, M.B.; validation, M.B. and M.D.; formal analysis, J.K.; data curation, M.L., J.K. and P.F.; writing—original draft preparation, M.D. and M.B.; writing—review and editing, M.D., P.F. and M.L.; visualization, P.F. and J.K. All authors have read and agreed to the published version of the manuscript.

Funding: This research was funded by project COGAR RFCR-CT-2013-00002, Slovak Research and Development Agency, under contract No. APVV-18-0526 and No. APVV-14-0892.

Institutional Review Board Statement: Not applicable.

Informed Consent Statement: Not applicable.

Data Availability Statement: No new data were created or analyzed in this study. Data sharing is not applicable to this article.

Acknowledgments: This work was supported by the Slovak Research and Development Agency under contract No. APVV-18-0526 and APVV-14-0892.

Conflicts of Interest: The authors declare no conflict of interest.

References

1. Bhutto, A.W.; Bazmi, A.A.; Zahedi, G. Underground coal gasification: From fundamentals to applications. *Prog. Energy Combust. Sci.* **2013**, *39*, 189–214. [CrossRef]
2. Uppal, A.A.; Bhatti, A.I.; Aamir, E.; Samar, R.; Khan, S.A. Control oriented modeling and optimization of one dimensional packed bed model of underground coal gasification. *J. Process. Control* **2014**, *24*, 269–277. [CrossRef]
3. Škvareková, E.; Tomašková, M.; Wittenberger, G.; Zeleňák, Š. Analysis of Risk Factors for Underground Coal Gasification. *Manag. Syst. Prod. Eng.* **2019**, *27*, 227–235. [CrossRef]
4. Xiao, Y.; Yin, H.; Duan, T.; Qi, H.; Zhang, Y.; Jolfaei, A.; Xia, K. An Intelligent prediction model for UCG state based on dual-source LSTM. *Int. J. Mach. Learn. Cybern.* **2020**. [CrossRef]
5. Canbaz, E.D.; Gur, M. Prediction of underground coal gasification performance of Turkish lignite reserves using stochiometric equilibrium model. *J. Therm. Sci. Technol.* **2020**, *40*, 195–205. [CrossRef]
6. Otto, C.; Kempka, T. Synthesis Gas Composition Prediction for Underground Coal Gasification Using a Thermochemical Equilibrium Modeling Approach. *Energies* **2020**, *13*, 1171. [CrossRef]
7. Janoszek, T.; Masny, W. CFD Simulations of Allothermal Steam Gasification Process for Hydrogen Production. *Energies* **2021**, *14*, 1532. [CrossRef]
8. Jowkar, A.; Sereshki, F.; Najafi, M. A new model for evaluation of cavity shape and volume during Underground Coal Gasification process. *Energy* **2018**, *148*, 756–765. [CrossRef]
9. Khan, M.M.; Mmbaga, J.; Shirazi, A.; Trivedi, J.; Liu, Q.; Gupta, R. Modelling Underground Coal Gasification—A Review. *Energies* **2015**, *8*, 12603–12668. [CrossRef]
10. Najafi, M.; Jalali, S.M.E.; KhaloKakaie, R.; Forouhandeh, F. Prediction of cavity growth rate during underground coal gasification using multiple regression analysis. *Int. J. Coal Sci. Technol.* **2015**, *2*, 318–324. [CrossRef]
11. Gur, M.; Canbaz, E.D. Analysis of syngas production and reaction zones in hydrogen oriented underground coal gasification. *Fuel* **2020**, *269*, 117331. [CrossRef]
12. Mandal, R.; Maity, T.; Chaulya, S.K.; Prasad, G.M. Laboratory investigation on underground coal gasification technique with real-time analysis. *Fuel* **2020**, *275*, 117865. [CrossRef]
13. Kapusta, K.; Wiatowski, M.; Stańczyk, K.; Zagorščak, R.; Thomas, H.R. Large-scale Experimental Investigations to Evaluate the Feasibility of Producing Methane-Rich Gas (SNG) through Underground Coal Gasification Process. Effect of Coal Rank and Gasification Pressure. *Energies* **2020**, *13*, 1334. [CrossRef]
14. Andrianopoulos, E.; Korre, A.; Durucan, S. Chemical Process Modelling of Underground Coal Gasification and Evaluation of Produced Gas Quality for End Use. *Energy Procedia* **2015**, *76*, 444–453. [CrossRef]
15. Yang, L. The Dynamic Temperature Field of Two-Stage Underground Coal Gasification (UCG). *Energy Sources Part A* **2006**, *28*, 667–680. [CrossRef]
16. Xin, L.; Wang, Z.; Huang, W.; Kang, G.; Lu, X.; Zhang, P.; Wang, J. Temperature field distribution of burnt surrounding rock in UCG stope. *Int. J. Min. Sci. Technol.* **2014**, *24*, 573–580. [CrossRef]
17. Kostúr, K.; Laciak, M.; Durdán, M. Some Influences of Underground Coal Gasification on the Environment. *Sustainability* **2018**, *10*, 1512. [CrossRef]
18. Luo, Y.; Margaretha, C.S.D. Comparison of UCG cavity growth with CFD model predictions. In Proceedings of the 7th International Conference on CFD in Minerals and Process Industries, Melbourne, Australia, 9–11 December 2009; CSIRO: Melbourne, Australia, 2009.
19. Janoszek, T.; Sygała, A.; Bukowska, M. CFD Simulation of Temperature Variation in Carboniferous Rock Strata during UCG. *J. Sustain. Min.* **2013**, *12*, 34–44. [CrossRef]
20. Otto, C.; Kempka, T. Thermo-mechanical Simulations Confirm: Temperature-dependent Mudrock Properties are Nice to have in Far-field Environmental Assessments of Underground Coal Gasification. *Energy Procedia* **2015**, *76*, 582–591. [CrossRef]
21. Kafle, R.C.; Pokhrel, K.P.; Khanal, N.; Tsokos, C.P. Differential equation model of carbon dioxide emission using functional linear regression. *J. Appl. Stat.* **2018**, *46*, 1246–1259. [CrossRef]

22. Motiee, H.; Ghasemnejad, S. Prediction of pipe failure rate in Tehran water distribution networks by applying regression models. *Water Supply* **2018**, *19*, 695–702. [CrossRef]
23. Hadavandi, E.; Hower, J.C.; Chelgani, S.C. Modeling of gross calorific value based on coal properties by support vector regression method. *Modeling Earth Syst. Environ.* **2017**, *3*, 37. [CrossRef]
24. Xu, L.; Cheng, Y.; Yin, R.; Zhang, Q. Comparative study of regression modeling methods for online coal calorific value prediction from flame radiation features. *Fuel* **2015**, *142*, 164–172. [CrossRef]
25. Gou, X.; Zhou, J.H.; Liu, J.Z.; Cen, K.F. Research on regression model of pulverized coal ignition temperature. *Energy Educ. Sci. Technol. Part A* **2011**, *28*, 143–150.
26. Chelgani, S.C.; Hower, J.C.; Jorjani, E.; Mesroghli, S.; Bagherieh, A.H. Prediction of coal grindability based on petrography, proximate and ultimate analysis using multiple regression and artificial neural network models. *Fuel Process. Technol.* **2008**, *89*, 13–20. [CrossRef]
27. Krzemień, A. Fire risk prevention in underground coal gasification (UCG) within active mines: Temperature forecast by means of MARS models. *Energy* **2019**, *170*, 777–790. [CrossRef]
28. Kapusta, K.; Stańczyk, K. Chemical and toxicological evaluation of underground coal gasification (UCG) effluents. The coal rank effect. *Ecotoxicol. Environ. Saf.* **2015**, *112*, 105–113. [CrossRef]
29. Shrestha, A.K.; Basnet, N. The Correlation and Regression Analysis of Physicochemical Parameters of River Water for the Evaluation of Percentage Contribution to Electrical Conductivity. *J. Chem.* **2018**, *2018*, 1–9. [CrossRef]
30. Huang, L.; Chen, J.C. A Multiple Regression Model to Predict In-process Surface Roughness in Turning Operation via Accelerometer. *J. Ind. Technol.* **2001**, *17*, 1–8.
31. Tavallali, P.; Razavi, M.; Brady, S. A non-linear data mining parameter selection algorithm for continuous variables. *PLoS ONE* **2017**, *12*, e0187676. [CrossRef] [PubMed]
32. Kačur, J.; Laciak, M.; Durdán, M.; Flegner, P. Utilization of Machine Learning Method. In Proceedings of the 18th International Carpathian Control Conference (ICCC), Sinaia, Romania, 28–31 May 2017.
33. Škvareková, E.; Taušová, M.; Seňová, A.; Wittenberger, G.; Novaková, J. Statistical Evaluation of Quantities Measured in the Detection of Soil Air Pollution of the Environmental Burden. *Appl. Sci.* **2021**, *11*, 3294. [CrossRef]
34. Bazaluk, O.; Lozynskyi, V.; Falshtynskyi, V.; Saik, P.; Dychkovskyi, R.; Cabana, E. Experimental Studies of the Effect of Design and Technological Solutions on the Intensification of an Underground Coal Gasification Process. *Energies* **2021**, *14*, 4369. [CrossRef]
35. Zagorščak, R.; Sadasivam, S.; Thomas, H.R.; Stańczyk, K.; Kapusta, K. Experimental study of underground coal gasification (UCG) of a high-rank coal using atmospheric and high-pressure conditions in an ex-situ reactor. *Fuel* **2020**, *270*, 117490. [CrossRef]
36. Durdán, M.; Terpák, J.; Kačur, J.; Laciak, M.; Flegner, P. Modeling of material balance from the experimental UCG. *Acta Polytech.* **2020**, *60*, 391–399. [CrossRef]
37. Aczel, A. *Complete Business Statistics*; Irwin: Homewood, IL, USA, 1989.
38. Montgomery, D. *Introduction to Statistical Quality Control*; Wiley: Hoboken, NJ, USA, 2009.
39. Rencher, A. *Methods of Multivariate Analysis*; Wiley: New York, NY, USA, 2002.
40. Hair, J.F.; Anderson, R.E.; Black, W.C.; Tatham, R.L. *Multivariate Data Analysis with Readings*; Prentice Hall: Englewood Cliffs, NJ, USA, 1995.

energies

MDPI

Article

Gas Permeability Model for Porous Materials from Underground Coal Gasification Technology

Grzegorz Wałowski [1,2]

[1] Institute of Technology and Life Sciences, National Research Institute, Falenty, 3 Hrabska Avenue, 05-090 Raszyn, Poland; g.walowski@itp.edu.pl
[2] Department of Renewable Energy, 67 Biskupinska Street, 60-463 Poznan, Poland

Abstract: Underground coal gasification (UCG) technology converts deep coal resources into synthesis gas for use in the production of electricity, fuels and chemicals. This study provides an overview of the systematic methods of the in situ coal gasification process. Furthermore, the model of the porous structure of coal has been presented and the gas movement taking place in the carbon matrix—which is part of the bed—has been described. The experimental tests were carried out with the use of air forced through the nozzle in the form of a gas stream spreading in many directions in a porous bed under bubbling conditions. The gas flow resistance coefficient was determined as a function of the Reynolds number in relation to the diameter of the gas flow nozzle. The proprietary calculation model was compared to the models of many researchers, indicating a characteristic trend of a decrease in the gas flow resistance coefficient with an increase in Reynolds number. The novelty of the study is the determination of the permeability characteristics of char (carbonizate) in situ in relation to melted waste rock in situ, taking into account the tortuosity and gas permeability factors for an irregularly shaped solid.

Keywords: underground coal gasification; georeactor; char; melted waste rock; gas permeability; tortuosity; porosity

Citation: Wałowski, G. Gas Permeability Model for Porous Materials from Underground Coal Gasification Technology. *Energies* **2021**, *14*, 4462. https://doi.org/10.3390/en14154462

Academic Editors: Marek Laciak, Ján Kačur and Milan Durdán

Received: 23 June 2021
Accepted: 21 July 2021
Published: 23 July 2021

1. Introduction

In situ thermal coal processing technology is currently a significant alternative to traditional coal gasification technologies in various aspects, both technical and technological. The in situ processing takes place in a natural deposit, which does not require the use of highly expensive and energy-consuming technological installations. Furthermore, as an UCG (underground coal gasification) process, it has the potential to obtain processed gas, related only to the scale of thermal gasification of the coal deposit. In both cases, the technology of in situ processing brings with it great production possibilities, also in the aspect of environmental protection and increasingly determined by unconventional techniques of processing minerals for energy purposes. The great advantage of UCG technology is that under the conditions of underground coal processing, all possible processes and reduction stages (obtaining syngas) are dealt with in one place and one bed, during which, after partial or complete gasification of the bed, a porous material in the form of a char is formed [1–3].

1.1. Systematics of Method of Gasification Process of Coal In Situ

It is possible to perform in-depth (30–200 m) processing, applying underground coal gasification (UCG) technology at a length of about 100 m (Table 1) [4].

In UCG technology, in fields excluded or for the farthest exploitation of uneconomic use, this method is applicable for the utilization of the most common shaft, excavations and mining pavements. Shallow boards applied in the case of this technique are streamlined (shallow deposits), opened or blind (on average shallow; Figure 1).

Table 1. Techniques for technology of processing coal in situ [4]. Reproduced from [4], the publisher: AGH University of Science and Technology 2012.

Underground Coal Gasification				
Method of processing	Process holes Injective (inflow)	Productive (outflow)	Channel Conective	Technique
Shaft	Vertical Vertical Horizontal Horizontal	Drooping Horizontal	Horizontal Point Vertical	Streamline Open Blind
Un-shaft	Directional Vertical Vertical (ignition)	Vertical Vertical	Pointing Horizontal	CRIP (I) CRIP (II) *"Knife edge"*

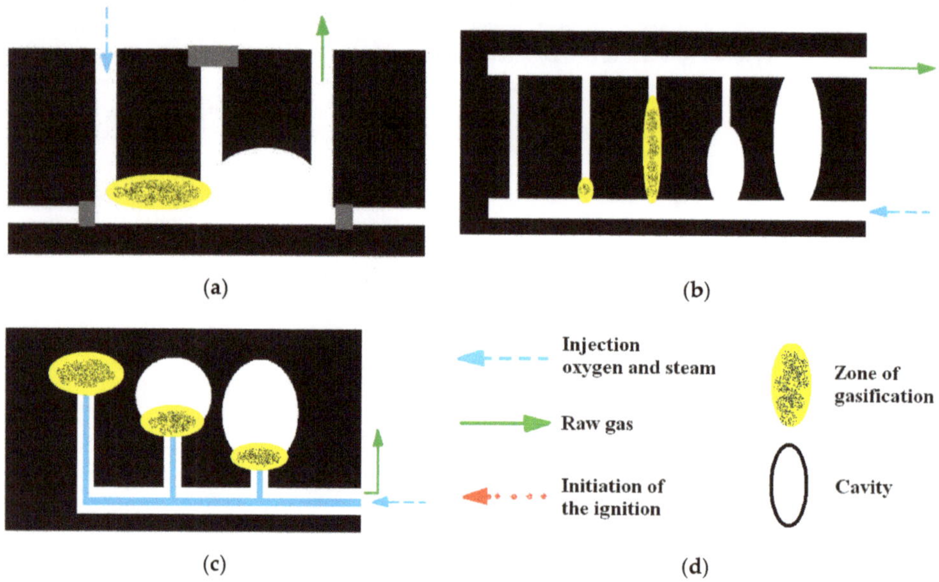

Figure 1. UCG shaft method: (**a**) streamlined, (**b**) open, (**c**) blind, (**d**) legend, acc. to [5]. Reproduced from [5], the publisher: Springer Nature Switzerland AG. 2011.

In this technology, a mixture of water, steam and air is pressed with a hole injection, below which gassification of the zone has been initiated. A receipt of raw gas follows on (step) from the contrarily placed productive hole. The georeactor is iniated in the course of this enduring process of freezing, which leads the geochamber in a result of the revolt cavern.

Performed analysis of the foundation of this technology indicates UCG technology, in that the shaft method presents the potential for coal seams which have not yet been exhausted in mines. An innovative project of the European Coal and Steel Community under the title "HUGE" (Hydrogen-oriented underground coal gasification for Europe) [6] is being fulfilled in Poland by the Main Institute of the Mining. The purpose of the project is the conduct of research over hydrogen production technology on underground coal gasification. Project "HUGE I" was brought to completion in the years 2007–2010. In carrying experimental research in the surface reactor ex situ [7], it conditions the processes in coal block gasification—Figure 2. It carries an attempt on a semi-technical scale in the Experimental Mine "Barbara"—in a coalfield, at a depth of 30 m at seam 310 (testing time of 16 days). The project "HUGE II" was completed in 2015, which includes structures of piloting installation localized in conditions in the coal mine stone mining seam 501 "Wieczorek".

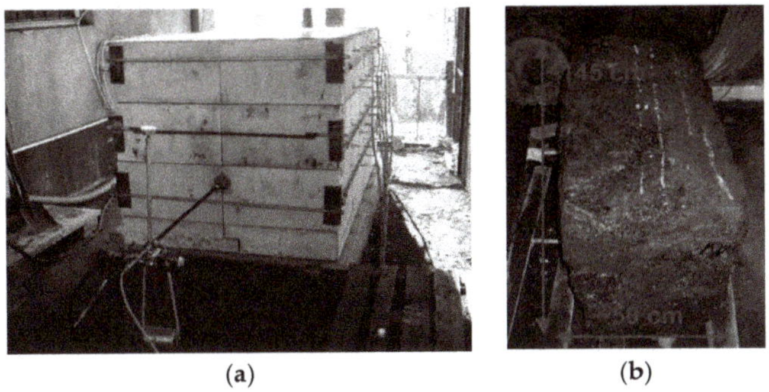

Figure 2. Experiment of the gasification coal: (**a**) reactor ex situ, (**b**) block of the hard bituminous coal, [6]. Reproduced from [6], the publisher: Publications Office of the European Union2012.

Next was the un-shaft method on the so-called CRIP technique (Controlled Retractable Injection Point). It takes advantage of the direction of boreholes in generation I in Figure 3a. This method presently developed into generation II of the step, of which characteristic features are presented in Figure 3b. It is an additional (third) hole example of injection-ignition in the process for the characteristic revolt of phenomenon underground gasification. It is described as "knife edge", which causes the efficient conversion of coal along with raw gas in front of the withdrawing fire [8].

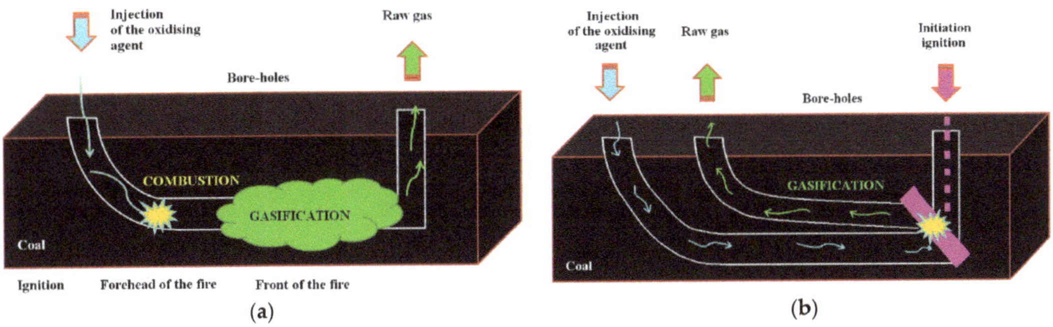

Figure 3. CRIP un-shaft method: (**a**) I—th generation, (**b**) II—second generation "Knife Edge", (own elaboration).

Numerous international experiences take advantage of the indicating methods of fabricating, within a range, the possession of underground coal gasification for commercial projects, for methods of producing syngas (Figure 4) [9]. However, arbitrated choice is a compound question under UCG process technology; it benefits to take into consideration the basic criteria of maximum energy proficiency within range.

Although it is possible to achieve the top depth of the field, at 2000 m it is possible to reach the top of this depth, along with creating, in this case, an exploitation field diameter of about 4000 m as in Complex Energy Extraction technology for Coal (CEEC) [8].

This technology takes advantage of the so-called Jet-Stinger (J-S) technique in operation with the method-based "Super Daisy Shaft" (SDS)—Figure 5. In essence, the system presents an armed main mineshaft SDS, in which, tube type direction boreholes are executed through the field under pressure drived in 8.1 MPa. Multiplicity of all of the system of a pipeline relies on a (J-S) concentric, with at least three tubes in the tube matching to create a"pipe in the pipe".

(**a**) (**b**) (**c**)

Figure 4. UCG trial: (**a**) Rocky Mountainn, Wyoming (USA), (**b**) Chinchila (Australia), (**c**) El Tremedal (Spain), [9]. Reproduced from [9], the publisher: CREATIVETIME 2012.

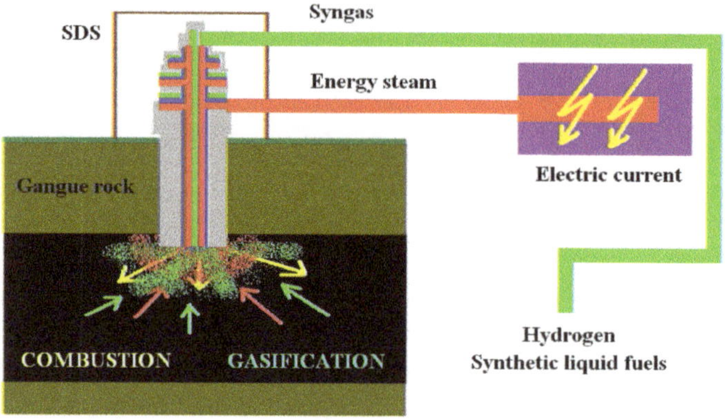

Figure 5. Method SDS of processing in georeactor, [10]. Reproduced from [10], the publisher: Poligraf 2011.

System multifunctionality of the pipe of the type (J-S) is shown in Figure 6. Among others it is related with, this is made possible through the induction of detonation propylants with a seismic wave. The agent performs the oxidizing injection to process coal, and to perform the pre-gasification and hydrogenation of the deposit. Moreover, the functional CO_2 sequestration, in receipt of procedural products, contains geothermal sources coming from liquid products which have a processing, as well as an overcrowding material to give a leg up for the filling-incurred cavern [10].

Figure 6. Part of the cycle of processing in SDS georeactor system, [10]. Reproduced from [10], the publisher: Poligraf 2011.

A technical foundation is presented for the described methods of underground coal gasification and decides the manner of translation of the process. However, they indicate the same range of application of UCG or CEEC technology. Utalitarian features of these technologies present big capabilities for practical application and for purposes of underground natural conversions of fuels, and for the considerable process range of raw gas.

1.2. Gas Movement in a Porous Structure

In this case, the hydrodynamic phenomena of the flow depend highly significantly on the pore structure as well as the forces and mechanisms causing the gas flow. As an example (Figure 7), following Seewald end Klein [11], it is possible to point to a diagram of the porous structure of carbon with the designation of the expected process mechanisms. In this case, the gas movement takes place in twisted and complicated microchannels—also in a system of interconnected channels with a different geometry. Filtration transport in a macropore system in such a structure [12] is associated with a significant modification of the pore pressure, which determines the course of diffusion in such a porous medium. In in situ conditions, the influence of external pressure (rock mass) and gas pressure may change the pore structure of the deposit. This is due to the fact that carbon can be treated as a biporous system, i.e., as microporous areas compressed by the high pressure of pore fluids (CO_2, CH_4, H_2O). As a consequence, the geometry of the macropores may change, or the macropores may become narrower due to the adsorption-absorption phenomena. Moreover, the structure of the medium significantly affects the permeability of the bed [11].

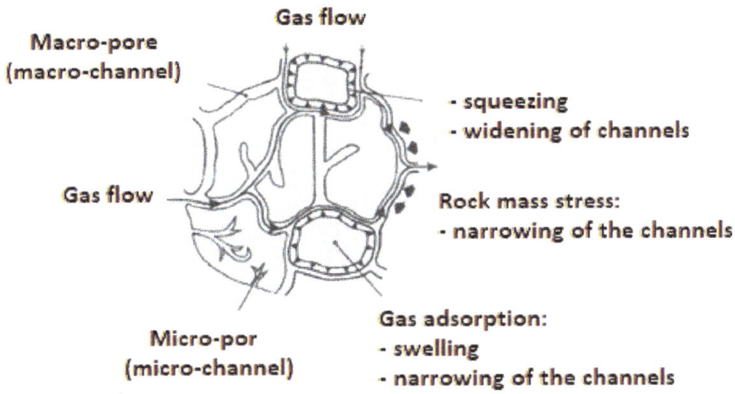

Figure 7. Model of the porous structure of coal according to Seewald, acc. to [11]. Reproduced from [11], the publisher: Gluckauf Forsch 1985.

The gas in a porous medium, due to the degree of its connection with the medium, can be divided into free gas (filling pores and crevices) and gas associated with the medium through sorption processes. Topolnicki et al. [12] state that from the physical point of view, a porous medium with submicro-, micro-, meso- and macropores can be treated as a sorbent with a structure in which the share of meso- and macropore surfaces is small. The porous bed system in which the gas flow takes place may be compact, uniform and containing fractures and fractures for which the permeability increases. Gas as a compressible medium occupies a volume dependent on pressure, and there is an equilibrium in the decompressed or disturbed seams. In an intact bed, free gas and adsorbed gas remain in equilibrium, while the violation of this equilibrium leads to the activation of gas filtration transport, mainly along the macropores. The outflow of gas from the fracture-porous medium reduces the pressure in the bed and is characterized by a specific bed permeability coefficient.

Several studies focused on in the fractured porous systems, the type of porosity systems is of significant importance for the gas flow in the porous structure [13–16]. The solutions proposed by the authors of these works are based on the following assumptions:

(1) For a single porous structure, the calculations are based on models of cylindrical capillaries:
   - Burdini—one capillary of equal radius;
   - Mualema—two bound capillaries with different radii;
   - The specific shape of the capillary in which the skin effect occurs.
(2) A double structure, in which there is a dominant gas flow through a porous matrix and fractures with negligible permeability.
(3) A multiple structure, in which a transient is created in the process of gas flow from the porous matrix to the fractures; these states determine the nature of the flow over relatively long periods of time (years).

In fact, gas flow in a coal seam occurs in a much more complex form that binds all of the above mechanisms together. This is due to the fact that in the process of gas release and flow in a fractured porous medium, the following phases of pressure reduction, desorption, diffusion and slotted flow can be distinguished. The phenomena of desorption and filtration are closely related in the mechanical and energetic sense [11]. The gas flow rate may also be determined by the desorption rate, independent of the permeability through the bed [17], for example, in Figure 8, the gas movement patterns resulting from the so-called rock matrix [18].

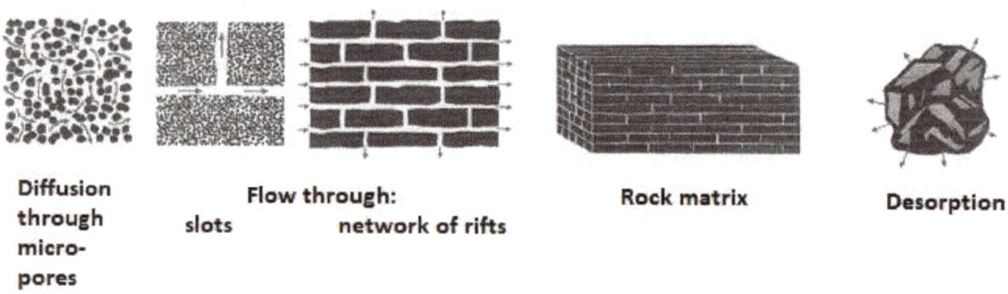

Figure 8. Models of gas movement in a fractured porous medium, acc. to [18]. Reproduced from [18], the publisher: Elsevier Science 1988.

Gas transport mechanisms that can only occur under specific pore geometry and under specific thermodynamic conditions are presented in Wyrwał [19].

This author distinguishes the movement of fluid in the bed according to the flow:

- Subcapillary—movement takes place only at increased temperature and pressure;
- Capillary—movement occurs under the influence of capillary forces and surface tension;
- Hypercapillary—motion according to the general laws of hydraulics and under the action of gravity.

On the other hand, Werner and Gertis [20] distinguish forms of transport as follows:

- Laminar flow occurs when the collisions of molecules occur in frequently changing capillary radii;
- Diffusion—occurs as a process of self-mixing of gas particles until full equalization;
- Knudsen transport—defined as the number of collisions of a molecule with the pore walls, related to the number of mutual collisions between molecules [21].

Considering the importance of coal porosity on syngas formation, the porosity and permeability of various ranks of coal seam with the coal bed methane (CBM) potential have been investigated widely [22–27].

In the literature, you can find a description of the fluid flow process in cylindrical capillaries, but the occurring phenomenon cannot be presented in the form of an unambiguous mathematical description. The main reason is the significant differentiation of the geometry and its variability in the longitudinal shape of individual pores, cross-section, connections between individual channels participating in the gas flow, etc.

Presented is a review of systematic methods of the coal gasification process in situ, thanks to which, concepts and tendencies for the use of unconventional clean energy technologies are determined. Such activities are aimed at rationalizing the use of fuels characterized by a model of porous-structure coal. An attempt was made to describe the gas movement taking place in the carbon matrix, which is part of the deposit.

Quite often, theoretical considerations of gas flow through a porous medium are based on models of flow through straight-axis channels (capillaries), flow laws—e.g., Poiseuille—and indications of flow resistance, often with some simplifying assumptions. In each case, however, these models are based on the quantities describing the physical properties of the porous medium. These include, among others, parameters such as: the shape of the pores, their size, interconnections, porosity of the system, both real and effective, characterizing the permeability of the porous bed.

With regard to the porous structure of coal and its chars considered in this study, the additional complexity of hydrodynamics results from the fact that the chars are skeletal structures, and therefore they are compact and in no way loosened during the increase in pressure in the system.

An original model of the total gas flow resistance through a porous medium was proposed, which takes into account the bed parameter related to the gas permeability coefficient and porosity.

1.3. Scope and Research Methodology

The aim of the research was to evaluate the gas permeability of materials with an irregular, fractured porous structure. The experimental tests were carried out with the use of air forced through the nozzle in the form of a gas stream spreading in many directions in a porous bed under bubbling conditions. The gas flow resistance coefficient was determined as a function of the Reynolds number, depending on the diameter of the gas flow nozzle. The total value of the gas permeability coefficient was determined experimentally, taking into account the authors' own model of the gas permeability coefficient and with the use of auxiliary functions in the context of tortuosity and porosity. As a consequence, the gas flow resistance coefficient as a function of the Reynolds number was determined for porous skeletal materials with bed velocity.

2. Materials and Methods

The study was based on various types of frame structures derived from UCG technology. Char (carbonizate) in situ—Figure 9a, and melted waste rock in situ—Figure 9b, was created as a result of thermal processing of hard coal in the deposit and comes from the "Barbara" Experimental Mine in Mikołów, Poland.

Research Position

The research was conducted on the laboratory position, as can be seen in Figure 10, the essential element of which was a vessel used to assess the phenomenon of aeration through a porous char material—Figure 11. The stand has been equipped with a rotameter for measuring the gas stream and a pressure-melted waste rock. The reference pressure related to the aeration process was determined with a reducer in the range (0.1–0.4) MPa.

Figure 11 shows the applied system of powering the sample for the free flow of gas (with emphasizing agreed parameters) and illustrates the flow of gas in these conditions.

The shape of the sample of this type along with the visible additive tube (see photographs) is showed in Figure 11—images refer to the volume sample of char (indefinable shape). The connection of the nozzle with the porous material is made through a hole, and then a special glue (binder) is applied—Figure 12.

(a)

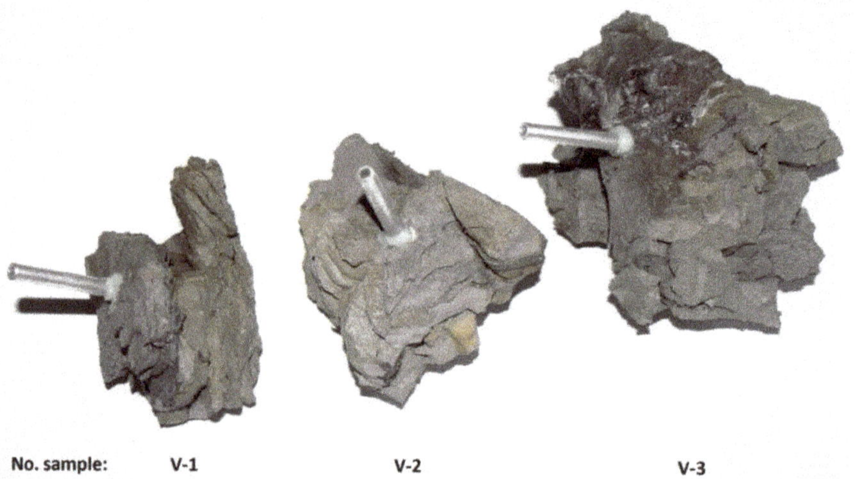

(b)

Figure 9. Research material samples: (**a**) char (carbonizate) in situ, (**b**) melted waste rock in situ (Photography by Grzegorz Wałowski).

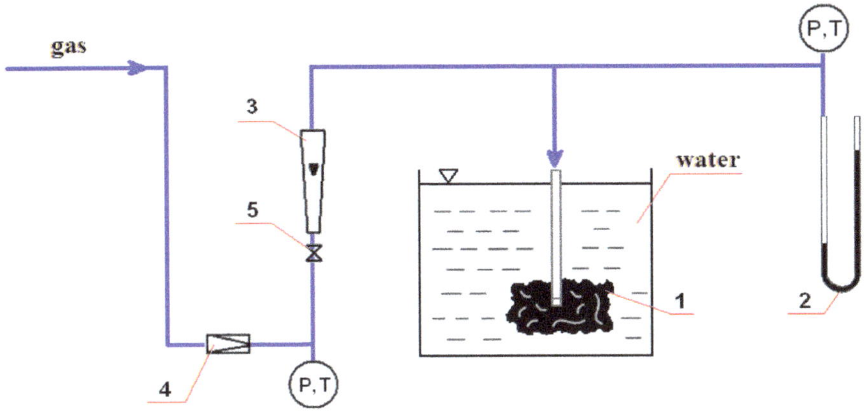

Figure 10. Scheme of the measuring system for testing the permeability of a porous material in bubble conditions (own elaboration): 1—porous material (sample), 2—differential pressure manometer, 3—rotameter (3 and bubble flow meter), 4—pressure reducers, 5—control valve, P—pressure gauge, T—thermometer.

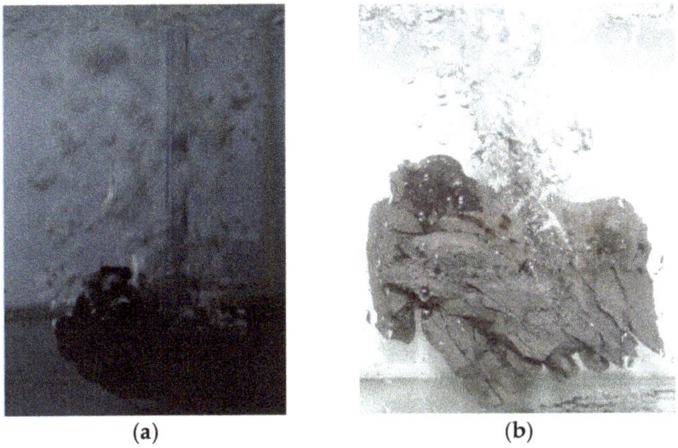

Figure 11. Flow of gas in bubbling conditions (Photo by Grzegorz Wałowski): (**a**) char (carbonizer) I-1; (**b**) melted waste rock V-3.

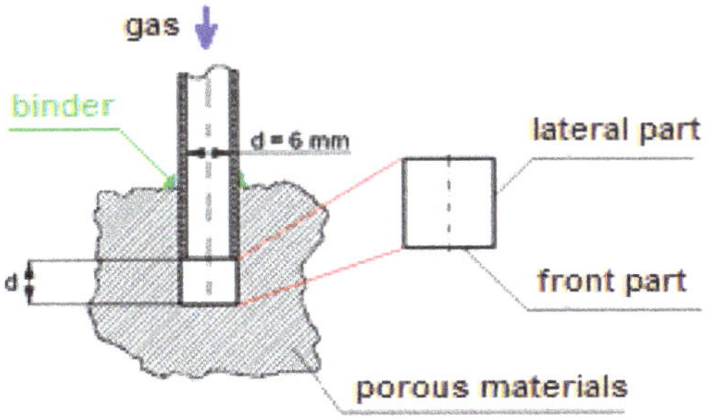

Figure 12. Volumetric sample feed system (own elaboration).

3. Results and Discussion

3.1. Results of Porosity and Gas Permeability

In quantification, the following parameters were assessed: porosity, porosity index and density—Table 2, permeability as a measure of pressure drop and a surrogate coefficient of flow resistance. Regardless of the measurement of the aeration flow, the permeability and the equivalent flow resistance coefficient were determined on the basis of the pressure drop across the bed of porous material. The determined parameters for the tested materials (samples) are presented in Table 3.

Table 2. Characteristics of the research material (own elaboration).

Material (Designation and Source Origin of Raw Material)		Porosity		Indicator Porosity	Density	
		Absolute	Effective		Apparent	Skeleton
Name	No. Sample	ε_b, %	ε_{ef}, %	e	ρ_a, kg/m^3	ρ_s, kg/m^3
char (carbonizer) in situ KD Barbara, Mikolow	I-1	42.2	21.1–33.7	0.7	1300	2250
	I-2	44.9	22.5–35.9	0.8	1239	
	I-3	33.9	17.0–27.1	0.5	1487	
	I-average	40.3	20.2–32.2	0.7	1342	2250
melted waste rock in situ KD Barbara, Mikolow	V-1	15.4	7.7–12.3	0.2	1438.4	1700
	V-2	36.4	18.2–29.1	0.6	1080.8	
	V-3	42.8	21.4–34.2	0.7	973.0	
	V-average	31.5	15.8–25.2	0.5	1164.1	1700

Table 3. Test results in conditions: air, 21.7 °C (own elaboration).

Research Material: Char (Carbonizer) in Itu KD Barbara, Mikolow No. Sample: I-1							
No.	Reference Pressure P_{re}, MPa	Gas Stream $Q_g \cdot 10^3$, m^3/s	Resistance Flow Measured ΔP_{zm}, kPa	No.	Reference Pressure P_{re}, MPa	Gas Stream $Q_g \cdot 10^3$, m^3/s	Resistance Flow Measured ΔP_{zm}, kPa
1	0.1	0.161	10.241	26	0.3	0.161	9.709
2	0.1	0.182	11.305	27	0.3	0.182	11.305
3	0.1	0.196	12.901	28	0.3	0.217	16.625
4	0.1	0.203	13.832	29	0.3	0.238	20.482
5	0.1	0.217	16.625	30	0.3	0.259	23.940
6	0.1	0.231	17.955	31	0.3	0.287	29.393
7	0.1	0.238	20.615	32	0.3	0.315	33.516
8	0.1	0.266	23.940	33	0.3	0.350	44.023
9	0.1	0.280	27.265	34	0.3	0.371	51.205
10	0.1	0.301	29.925	35	0.3	-	-
11	0.1	0.329	37.905	36	0.3	-	-
12	0.1	0.350	42.826	37	0.3	-	-
13	0.1	0.371	49.210	38	0.3	-	-
14	0.2	0.161	9.75	39	0.4	0.161	10.640
15	0.2	0.189	12.635	40	0.4	0.196	12.768
16	0.2	0.210	15.295	41	0.4	0.231	19.285
17	0.2	0.231	19.285	42	0.4	0.266	24.206
18	0.2	0.252	23.275	43	0.4	0.301	31.255
19	0.2	0.280	26.999	44	0.4	0.336	39.900
20	0.2	0.301	31.255	45	0.4	0.350	43.624
21	0.2	0.322	36.176	46	0.4	0.371	51.205
22	0.2	0.336	38.969	47	0.4	-	-
23	0.2	0.343	42.826	48	0.4	-	-
24	0.2	0.371	49.476	49	0.4	-	-
25	0.2	0.392	53.466	50	0.4	-	-

Table 3. *Cont.*

Research material: Char (carbonizer) in situ KD Barbara, Mikolow No. Sample: I-2							
No.	**Reference Pressure P_{re}, MPa**	**Gas Stream $Q_g \cdot 10^3$, m^3/s**	**Resistance Flow Measured ΔP_{zm}, kPa**	**No.**	**Reference Pressure P_{re}, MPa**	**Gas Stream $Q_g \cdot 10^3$, m^3/s**	**Resistance Flow Measured ΔP_{zm}, kPa**
1	0.1	0.161	7.448	26	0.3	0.378	28.196
2	0.1	0.182	7.315	27	0.3	0.427	30.324
3	0.1	0.196	8.512	28	0.3	0.539	46.683
4	0.1	0.231	11.305	29	0.3	0.553	49.210
5	0.1	0.252	12.236	30	0.3	0.637	66.633
6	0.1	0.266	13.965	31	0.3	0.686	85.386
7	0.1	0.301	15.960	32	0.3	-	-
8	0.1	0.322	17.955	33	0.3	-	-
9	0.1	0.357	20.349	34	0.3	-	-
10	0.1	0.385	24.472	35	0.3	-	-
11	0.1	0.427	31.654	36	0.3	-	-
12	0.1	-	-	37	0.3	-	-
13	0.1	-	-	38	0.3	-	-
14	0.2	0.161	6.916	39	0.4	0.532	45.486
15	0.2	0.196	8.246	40	0.4	0.637	67.830
16	0.2	0.231	9.975	41	0.4	0.651	73.283
17	0.2	0.252	12.236	42	0.4	0.728	99.085
18	0.2	0.280	13.965	43	0.4	0.763	114.513
19	0.2	0.308	16.226	44	0.4	-	-
20	0.2	0.336	19.285	45	0.4	-	-
21	0.2	0.357	20.881	46	0.4	-	-
22	0.2	0.385	24.605	47	0.4	-	-
23	0.2	0.413	28.595	48	0.4	-	-
24	0.2	0.490	39.900	49	0.4	-	-
25	0.2	0.553	49.343	50	0.4	-	-
Research material: Char (carbonizer) in situ KD Barbara, Mikolow No. Sample: I-3							
No.	**Reference Pressure P_{re}, MPa**	**Gas Stream $Q_g \cdot 10^3$, m^3/s**	**Resistance Flow Measured ΔP_{zm}, kPa**	**No.**	**Reference Pressure P_{re}, MPa**	**Gas Stream $Q_g \cdot 10^3$, m^3/s**	**Resistance Flow Measured ΔP_{zm}, kPa**
1	0.1	0.161	8.911	26	0.3	0.308	30.856
2	0.1	0.196	12.635	27	0.3	0.434	65.436
3	0.1	0.231	17.556	28	0.3	0.490	86.982
4	0.1	0.266	21.546	29	0.3	0.546	108.661
5	0.1	0.301	27.664	30	0.3	0.588	128.079
6	0.1	-	-	31	0.3	-	-
7	0.1	-	-	32	0.3	-	-
8	0.1	-	-	33	0.3	-	-
9	0.1	-	-	34	0.3	-	-
10	0.1	-	-	35	0.3	-	-
11	0.1	-	-	36	0.3	-	-
12	0.1	-	-	37	0.3	-	-
13	0.1	-	-	38	0.3	-	-
14	0.2	0.161	9.709	39	0.4	0.371	48.146
15	0.2	0.196	13.433	40	0.4	0.525	99.085
16	0.2	0.238	18.354	41	0.4	0.581	126.350
17	0.2	0.343	38.703	42	0.4	0.623	140.980
18	0.2	0.385	50.939	43	0.4	-	-
19	0.2	0.434	64.505	44	0.4	-	-
20	0.2	0.476	77.672	45	0.4	-	-
21	0.2	-	-	46	0.4	-	-
22	0.2	-	-	47	0.4	-	-
23	0.2	-	-	48	0.4	-	-

Table 3. *Cont.*

24	0.2	-	-	49	0.4	-	-
25	0.2	-	-	50	0.4	-	-

Research material: Melted waste rock in situ KD Barbara, Mikolow
No. Sample: V-1

No.	Reference Pressure P_{re}, MPa	Gas Stream $Q_g \cdot 10^3$, m^3/s	Resistance Flow Measured ΔP_{zm}, kPa	No.	Reference Pressure P_{re}, MPa	Gas Stream $Q_g \cdot 10^3$, m^3/s	Resistance Flow Measured ΔP_{zm}, kPa
1	0.1	0.161	2.660	26	0.3	0.161	2.660
2	0.1	0.196	3.325	27	0.3	0.196	3.325
3	0.1	0.231	3.724	28	0.3	0.231	3.990
4	0.1	0.266	4.655	29	0.3	0.266	4.655
5	0.1	0.301	5.320	30	0.3	0.301	5.320
6	0.1	0.336	5.985	31	0.3	0.336	5.985
7	0.1	0.371	6.650	32	0.3	0.371	7.315
8	0.1	0.406	7.315	33	0.3	0.406	7.980
9	0.1	0.441	8.246	34	0.3	0.441	9.310
10	0.1	0.476	9.310	35	0.3	0.476	10.374
11	0.1	0.511	9.975	36	0.3	0.511	11.970
12	0.1	0.546	10.640	37	0.3	0.546	12.635
13	0.1	-	-	38	0.3	-	-
14	0.2	0.161	2.660	39	0.4	0.161	2.660
15	0.2	0.196	3.325	40	0.4	0.196	3.325
16	0.2	0.231	3.990	41	0.4	0.231	3.990
17	0.2	0.266	4.655	42	0.4	0.266	4.655
18	0.2	0.301	5.320	43	0.4	0.301	5.320
19	0.2	0.336	5.985	44	0.4	0.336	5.985
20	0.2	0.371	6.650	45	0.4	0.371	7.315
21	0.2	0.406	7.980	46	0.4	0.406	7.980
22	0.2	0.441	8.645	47	0.4	0.441	9.310
23	0.2	0.476	9.975	48	0.4	0.476	9.975
24	0.2	0.511	11.305	49	0.4	0.511	11.305
25	0.2	0.546	11.970	50	0.4	0.546	11.970

Research material: Melted waste rock in situ KD Barbara, Mikolow
No. Sample: V-2

No.	Reference Pressure P_{re}, MPa	Gas Stream $Q_g \cdot 10^3$, m^3/s	Resistance Flow Measured ΔP_{zm}, kPa	No.	Reference Pressure P_{re}, MPa	Gas Stream $Q_g \cdot 10^3$, m^3/s	Resistance Flow Measured ΔP_{zm}, kPa
1	0.1	0.161	7.315	26	0.3	0.161	7.448
2	0.1	0.196	9.975	27	0.3	0.196	10.241
3	0.1	0.231	13.300	28	0.3	0.231	13.965
4	0.1	0.266	16.891	29	0.3	0.266	18.221
5	0.1	0.301	20.216	30	0.3	0.301	21.945
6	0.1	0.336	25.403	31	0.3	0.336	26.866
7	0.1	0.371	32.58.	32	0.3	0.371	31.255
8	0.1	0.406	37.95	33	0.3	0.406	38.703
9	0.1	0.441	44.55	34	0.3	0.441	45.885
10	0.1	0.476	50.540	35	0.3	0.476	52.136
11	0.1	0.511	55.195	36	0.3	0.511	58.520
12	0.1	0.539	59.185	37	0.3	0.532	62.909
13	0.1	-	-	38	0.3	-	-
14	0.2	0.161	7.315	39	0.4	0.161	7.315
15	0.2	0.196	10.640	40	0.4	0.196	10.241
16	0.2	0.231	13.300	41	0.4	0.231	14.231
17	0.2	0.266	16.625	42	0.4	0.266	17.955
18	0.2	0.301	20.615	43	0.4	0.301	21.945
19	0.2	0.336	25.935	44	0.4	0.336	27.265
20	0.2	0.371	30.324	45	0.4	0.371	32.585

Table 3. *Cont.*

21	0.2	0.406	37.240	46	0.4	0.406	39.235
22	0.2	0.441	44.023	47	0.4	0.441	46.550
23	0.2	0.476	49.875	48	0.4	0.476	51.870
24	0.2	0.511	60.116	49	0.4	0.511	61.845
25	0.2	0.546	62.510	50	0.4	0.546	62.510

Research material: Melted waste rock in situ KD Barbara, Mikolow
No. Sample: V-3

No.	Reference Pressure P_{re}, MPa	Gas Stream $Q_g \cdot 10^3$, m^3/s	Resistance Flow Measured ΔP_{zm}, kPa	No.	Reference Pressure P_{re}, MPa	Gas Stream $Q_g \cdot 10^3$, m^3/s	Resistance Flow Measured ΔP_{zm}, kPa
1	0.1	0.161	5.586	26	0.3	0.161	5.985
2	0.1	0.196	8.113	27	0.3	0.196	8.246
3	0.1	0.231	11.172	28	0.3	0.231	11.305
4	0.1	0.266	13.699	29	0.3	0.266	14.098
5	0.1	0.301	16.891	30	0.3	0.301	17.423
6	0.1	0.336	21.014	31	0.3	0.336	21.546
7	0.1	0.371	24.472	32	0.3	0.371	27.398
8	0.1	0.406	28.595	33	0.3	0.406	32.585
9	0.1	0.441	35.245	34	0.3	0.441	38.171
10	0.1	0.476	41.496	35	0.3	0.476	45.486
11	0.1	0.511	47.215	36	0.3	0.511	49.875
12	0.1	0.532	54.530	37	0.3	0.539	53.466
13	0.1	–	–	38	0.3	–	–
14	0.2	0.161	5.719	39	0.4	0.161	5.320
15	0.2	0.196	8.246	40	0.4	0.196	7.980
16	0.2	0.231	11.305	41	0.4	0.231	11.305
17	0.2	0.266	13.965	42	0.4	0.266	13.965
18	0.2	0.301	16.625	43	0.4	0.301	17.024
19	0.2	0.336	20.615	44	0.4	0.336	21.280
20	0.2	0.371	25.270	45	0.4	0.371	26.600
21	0.2	0.406	29.925	46	0.4	0.406	31.521
22	0.2	0.441	36.841	47	0.4	0.441	37.905
23	0.2	0.476	43.225	48	0.4	0.476	43.225
24	0.2	0.511	48.013	49	0.4	0.511	49.476
25	0.2	0.546	50.806	50	0.4	0.546	55.594

In order to achieve the aim of the study, detailed experimental studies were carried out to assess gas permeability in the structure of the porous material, and the results are presented in Table 3.

3.2. Results of Coefficient of Gas Flow Resistances

The results for the determined flow resistance coefficient Equation (1)

$$\xi_r = \frac{2}{\rho_g w_r^2} \Delta P_{zm} \tag{1}$$

for the volumetric sample are given in Figure 13.

The reference of the value of this coefficient to the Reynolds number Equation (2) was used at a gas speed of w_o resulting from the d diameter of the feeding nozzle—Figure 12.

$$Re_r = \frac{w_r d_r \rho_g}{\eta_g} \tag{2}$$

These results indicate a decrease in changes in the value of the drag coefficient as a result of an increase in the Reynolds number, which is consistent with the physics of the analyzed phenomena, but the scale of these changes is sometimes extremely large. This proves that the flow resistance is highly influenced by the dynamics of gas flow through the porous material, in particular, the disturbance of the velocity profile. This trend and measuring range both indicate that for coal char and melted waste rock (Figure 13a), the turbulent nature of the gas movement is noted, which is proven by the non-linear nature of this coefficient. It may be noted that melted waste rock has the minimum flow resistance, relative to coal char (Figure 13). Undoubtedly, this is due to the fact that this material—with a low porosity (average 31.5%)—has a highly extensive system of pores and channels closed to gas flow. On the other hand, the char with higher porosity (average 40.3%) has a highly extensive system of pores and open channels for gas flow. Figure 13 proves the change in the value of the drag coefficient proportional to the Reynolds number, that when modeling the hydrodynamic conditions of gas flow through the porous material of the skeleton, the relationship is as follows Equation (3):

$$\xi_{\varepsilon} = f(\mathrm{Re},\ \varepsilon) \tag{3}$$

This also applies to the Reynolds number, which in this case may take a different form (Table 4).

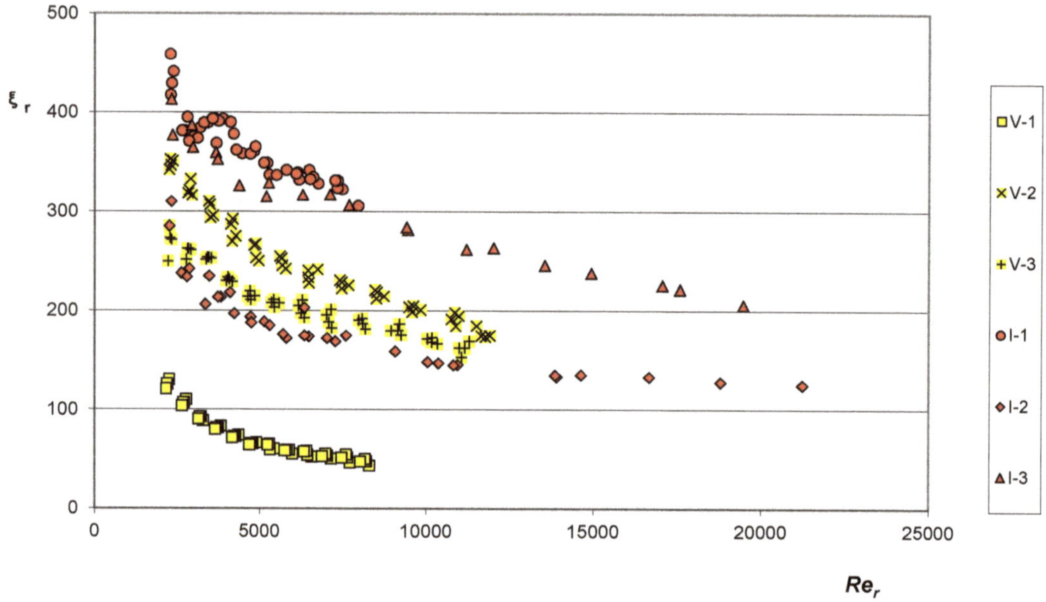

Figure 13. Coefficient of gas flow resistances for volume sample (own elaboration): coal char (carbonizer) in situ: I-1, I-2, I-3; melted waste rock in situ: V-1, V-2, V-3.

The presented exemplary gas flow characteristic for char (carbonize) in Figure 14 indicates a discrepancy in relation to other calculation models included in Table 4. Figure 14 relates to the Reynolds number depending on the substitute coefficient of flow resistance calculated for the full flow of the supply nozzle. The discrepancies may result from the lack of identification of model solutions for the specific structure of the tested materials. The structure of these materials shows the features of numerous closed and blind pores for gas flow. This indicates a diversified permeability and high randomness of the structure of this type of materials, depending on the type and conditions of in situ thermal processing of coal. When interpreting Figure 14, it should be noted that the mean equivalent flow resistance coefficient does not depend on the shape of the solid, but on the internal structure, as shown by the distribution of experimental points in Figure 14 for all analyzed models. In Table 4, the equations are not adequate to describe the hydrodynamics of the tested materials. The results of the comparison of the models according to the authors show that the possible adaptation of the computational methods (models-coefficient of resistances of the flow through granular porous structures) characteristic for porous granular beds does not give sufficient results for the use of these models in the description of the gas flow hydrodynamic criteria. The main reason for such a situation is the influence of pressure losses on the bed of the tested material, related to friction losses and losses resulting from the disturbance of the velocity profile—especially the local resistance of the skeleton material.

Table 4. Correlation equations for calculating the coefficient of flow resistance through granular porous structures (own elaboration).

Autor	Model Equation		Criteria Number	
Ergun [28]	$\zeta_{rE} = \frac{150}{Re_\varepsilon} + 1.75$	(4)	$Re_r = \frac{w_r d_r \rho}{(1-\varepsilon)\eta}$	(5)
Brauer [29]	$\zeta_{rB} = \frac{160}{Re_\varepsilon} + \frac{3.1}{Re_\varepsilon^{0.1}}$	(6)		
Tallmadge [30]	$\zeta_{rT} = \frac{150}{Re_\varepsilon} + \frac{4.2}{Re_\varepsilon^{0.1666}}$	(7)		
Burke-Plummer [31]	$\zeta_{rB-P} = 0.878\frac{(1-\varepsilon)}{\varepsilon^2}$	(8)		
Blake-Kozeny [32]	$\zeta_{rB-K} = 75\frac{(1-\varepsilon)^2}{\varepsilon^3}\frac{1}{Re_\varepsilon}$	(9)	$Re_r = \frac{w_r d_r \rho}{\eta}$	(10)
Blake-Kozeny–Carman [32]	$\zeta_{rB-K-C} = \frac{180}{Re_\varepsilon}$	(11)		
Żaworonkow [33]	$\zeta_{rZ} = \frac{3.8}{Re_\varepsilon^{0.2}}$	(12)	$Re_r = \frac{w_r d_r \rho}{\varepsilon\eta}$	(13)
Windsperger [34]	$\zeta_{rW} = 2.2\left(\frac{0.4}{\varepsilon}\right)^{0.78}\left(\frac{64}{Re_\varepsilon} + \frac{1.8}{Re_\varepsilon^{0.1}}\right)$	(14)	$Re_r = \frac{2}{3}\frac{w_r d_r \rho}{(1-\varepsilon)\eta}$	(15)

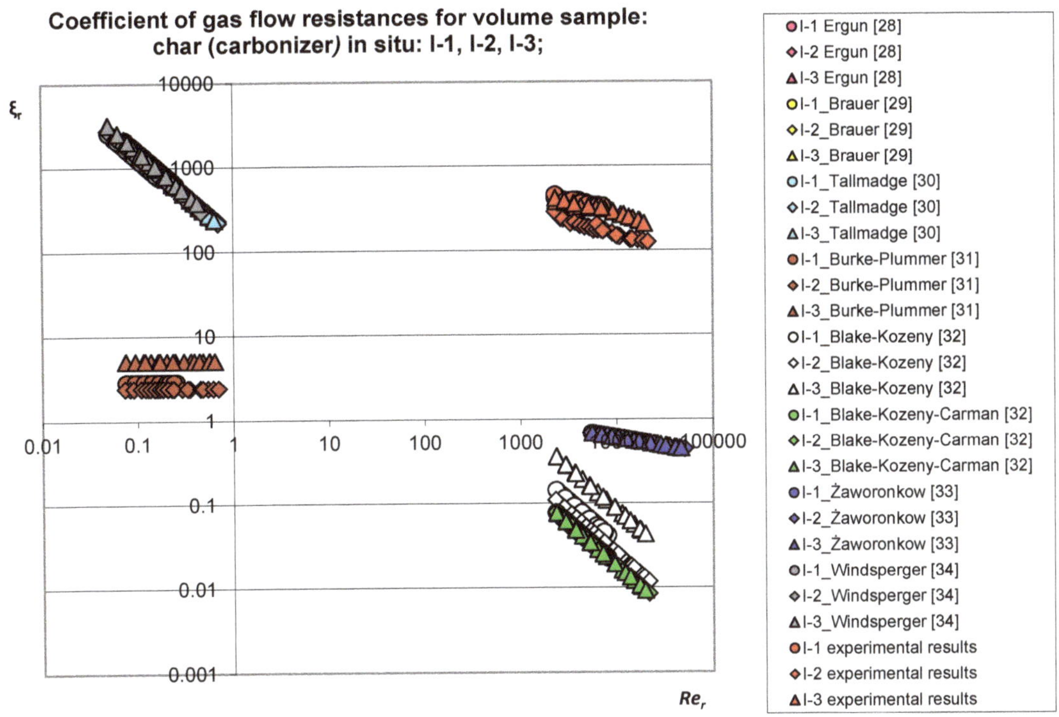

Figure 14. Correlation referred to other authors in acc. with Table 4 a coefficient of resistances for volume sample char.

There is still a debate [35] on how to best describe this criteria number to identify the flow through frame-structured porous materials. In the case of these materials, it is extremely difficult or even impossible to assess the diameters of pores and capillaries and their actual flow velocity; Bear and Cheng [36] suggest that, in this case, the Reynolds number Equation (16) to be determined for the total volume of the porous material is related to the flow path:

$$Re_{\varepsilon} = \frac{w_{\varepsilon} d_{\varepsilon}^{*} \rho_{g}}{\eta_{g}} \quad (16)$$

The characteristic linear measurement is calculated as an alternative diameter, resulting from the volume of the porous material and the section active for the flow, viz. Equation (17):

$$d_{\varepsilon}^{*} = \frac{V_c}{\varepsilon A_r} \quad (17)$$

Furthermore, velocity is a result of the deposit porosity and is associated with an apparent velocity (calculated for the entire cross-section of the deposit; Equation (18)):

$$w_{\varepsilon} = \varepsilon w_r \quad (18)$$

The conducted analyses show that the method resulting from Equations (16)–(18), and the definition of the criteria, the Reynolds number, does not best reflect any of the hydrodynamic conditions in which a low porous frame material gas is present. This is due to the fact that the subject of the research was a sample of a different shape. Particularly in the case of the volume of a solid, the determination of the cross-sections for the direction of flow is extremely difficult and imprecise. In order to solve these problems, an attempt was made to develop an alternative model based on the change of kinetic energy characteristics for all gas flow oppositions through the porous medium. According to Equation (1), this may be as follows Equation (19):

$$\zeta_{\varepsilon}(\mathrm{Re}) = \frac{2\Delta P_c}{\rho_g w_{\varepsilon}^{2}} \quad (19)$$

3.3. *Results of Gas-Permeability Coefficient*

When pointing to the problem, the diversified shape of the material and the characteristic structural features resulting from its porosity and permeability were taken into account. Equation (19) can be modified by introducing a correction coefficient in the form of the so-called tortuosity parameter Equation (20):

$$\Psi_{\varepsilon} = f(K_V^{*}, \varepsilon) \quad (20)$$

Assuming the gas flow rate, pressure drop across the bed, porosity of the bed and the type of gas, the value of the gas permeability coefficient Equation (21) can be determined experimentally and determined [37]:

$$K_V^{*} = \frac{Q_g}{\sqrt{\frac{\Delta P_{zm}}{\rho_g}}} \quad (21)$$

Results of the measurement being characteristic of a permeability coefficient of the gas of the tested sample of char were shown on Figure 15.

The measured gas flow through the char from the total pressure drop is greater the higher the aeration pressure, i.e., the reference pressure. On the other hand, the carbon permeability is constantly increasing by the value of a given Reynolds number.

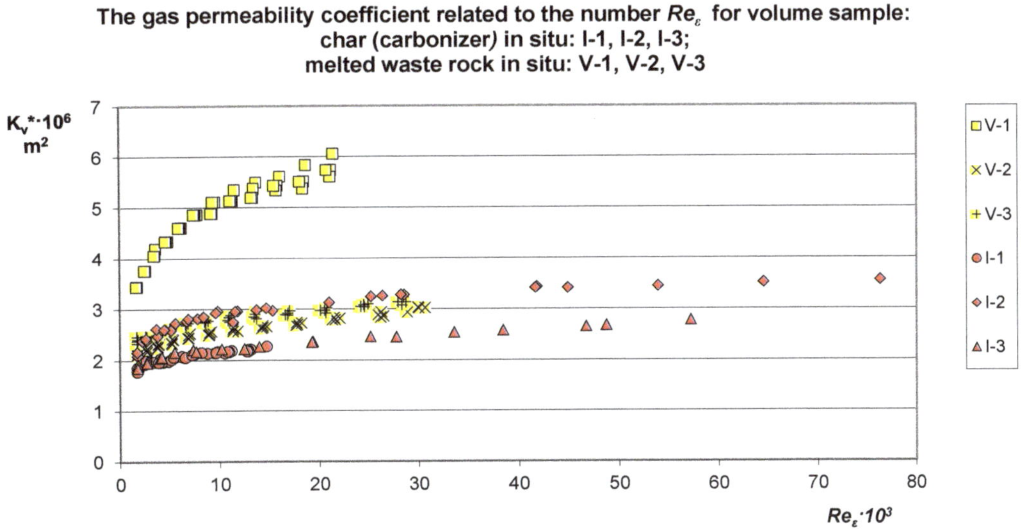

Figure 15. Gas-permeability coefficient (own elaboration): char (carbonizer) in situ: I-1, I-2, I-3; melted waste rock in situ: V-1, V-2, V-3.

3.4. Own Model of Coefficient of Gas Flow Resistances

Taking into account dependence Equation (20), the equation for the total coefficient of resistance is as follows Equation (22):

$$\xi_c(Re_\varepsilon^*) = \frac{2\Delta P_c}{\rho_g w_\varepsilon^2}\Psi_\varepsilon \quad (22)$$

With reference to Equation (22), compensation calculations were performed for a sample in the form of a solid body of a different shape (volume). The arithmetic analysis shows that for a volumetric solid, the tortuosity parameter should be calculated on the basis of the relationship Equation (23):

$$\Psi_\varepsilon = \frac{\chi_\varepsilon^a}{Re_\varepsilon^*} \quad (23)$$

The auxiliary function for the exponent: a is −0.22 for char and a is −0.20 for melted waste rock; at the base of power is the coefficient of bed formation (24) related to the gas permeability coefficient and porosity:

$$\chi_\varepsilon = K_V^{*(\varepsilon-1)} \quad (24)$$

While the Reynolds number Equation (25) in the auxiliary function for the exponent b is 0.4, char and melted waste rock take into account the defined apparent velocity Equation (26):

$$Re_\varepsilon^* = \left(\frac{w_\varepsilon^* d_r \rho_g}{\eta_g}\right)^b \quad (25)$$

The apparent velocity Equation (26) refers to the entire space of the volume sample feeding as the cross-section resulting from the frontal and lateral area of the feeding nozzle (Figure 12).

$$w_\varepsilon^* = \frac{Q_g}{A_o^*} \quad (26)$$

Cross-sectional area of nozzle feeding the porous material Equation (27):

$$A_o^* = \frac{5}{4}\varepsilon_b \pi d_r^2 \quad (27)$$

Based on our own model, the obtained results of the resistance coefficient function as a function of the Reynolds number are shown in Figure 16.

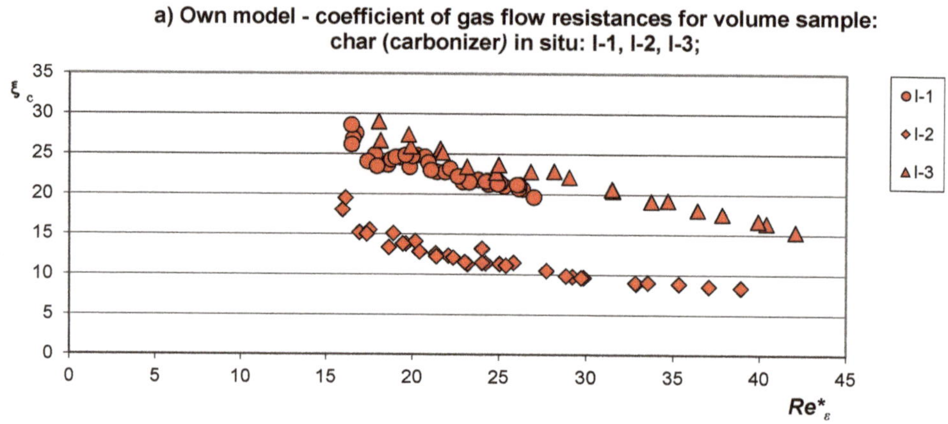

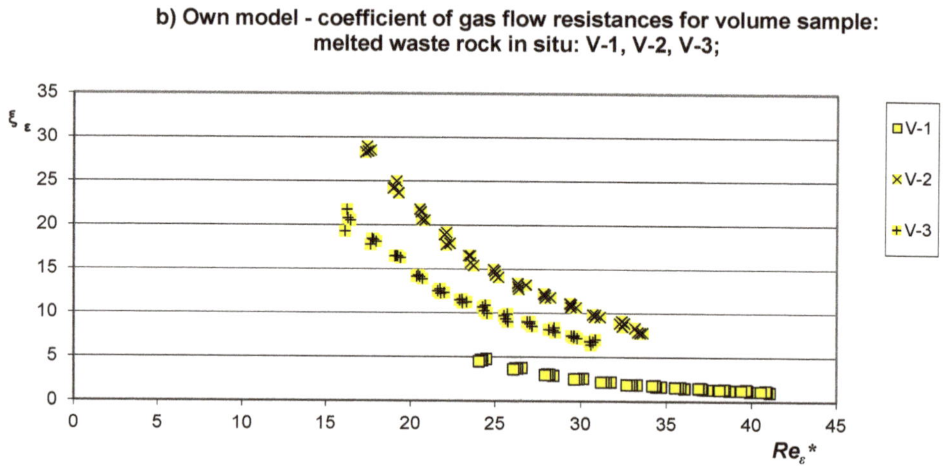

Figure 16. Coefficient of resistances of gas flow through: (**a**) char in situ; (**b**) melted waste rock in situ—own models (own elaboration).

The dependence presented in Figure 16, corrected by parameters resulting from the authors' own research, was compared to the points resulting from other calculation models [28–34]. In this way, the developed area was obtained in line with the porous sediment permeability trend–char in situ is presented as an example Figure 17.

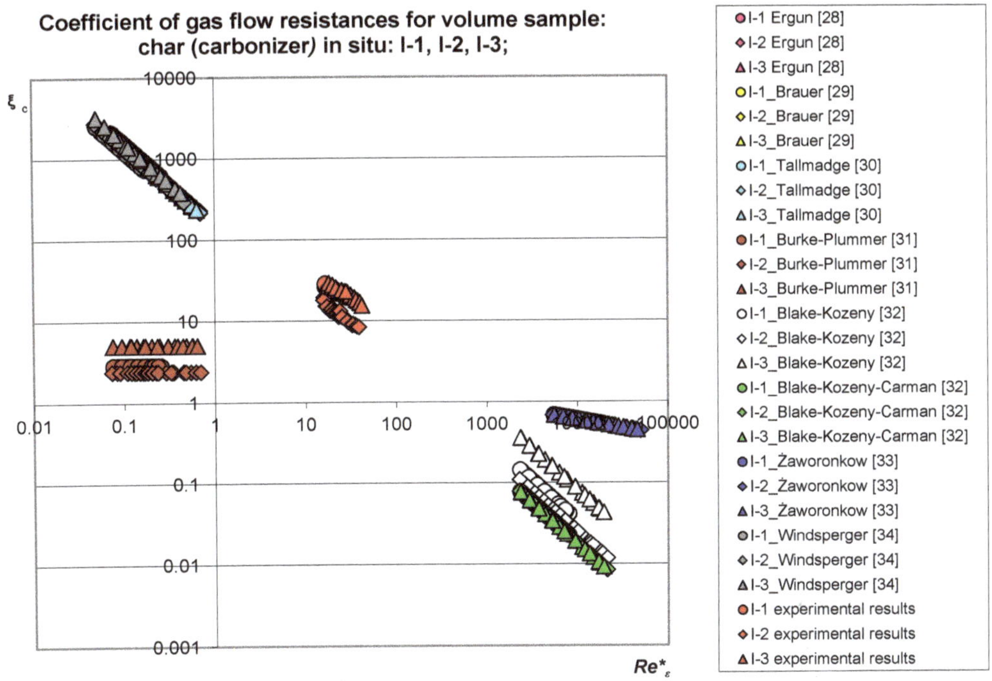

Figure 17. Coefficient of resistances of gas flow through coal char acc. Authors' models (Table 4)—the reference of the authors' own model to the computational models of other authors (own elaboration).

The distribution of experimental points proves that the expectation for the proprietary model for all other models, according to the authors (Table 4), shows the same trend of changes in the total flow resistance coefficient in the relation of the modified Reynolds number, which confirms the adequacy of the assumptions. It should be noted that for the methodology adopted in this way for the use of the authors' own model, it is possible to develop an area that is not yet recognized, especially in the context of gas flow under UCG conditions.

In this research, direct measurements were taken, and for this purpose, instruments for measuring gas flow, pressure (pressure difference) and temperature were used. These instruments have been properly calibrated, for which the results in relation to the gas flow meters used are given in Table 5.

Table 5. Parameters of gas flow meters calibration (own elaboration).

Flow Meter Type	Measurement Range	Scaling Equation—The Value of the Air Stream, dm^3/min	Accuracy of Scaling
RDN 06–03	0–1.9	Q_g = (0.0137 scala) − 0.30086	0.97
R 10a	0–38	Q_g = (0.2836 scala) + 9.9091	0.99
RDN 06–03	0–48	Q_g = (0.216 · scala) + 1.4112	0.99
R 10m	0–51	Q_g = (0.4264 scala) + 9.5	0.99
R	0–1.5	Q_g = (10 mL/ measurement time)	±5%

The subject of my own considerations was also the analysis of the measurement error. It was found that in relation to the measured values, this analysis did not bring a significant improvement in the estimation of the measurement error. Therefore, a detailed description of the measurement error analysis was omitted.

As an example, the algorithm for calculating the error and uncertainty of the air volume flow measurement is presented, which results from the analysis of measurement errors. The results of the calculations according to the algorithm are presented in Table 6.

Table 6. An algorithm concerning the analysis of measurement error (own elaboration).

Algorithm	Score
Expected value as an arithmetic mean.	3.74×10^{-4}
Measure of dispersion as the variance of the arithmetic mean.	1.07×10^{-10}
Standard deviation.	1.03×10^{-5}
The component of the measurement error limit: — systematic limit error, where the absolute error of the measuring instrument (rotameter RDN06–03) is 5%.	1.87×10^{-6}
The component of the measurement error limit: — accidental random error.	3.01×10^{-5}
Measurement error limit at the probability confidence level $p \sim 0.99$.	3.29×10^{-5}
The measurement result at the confidence level $p \sim 0.99$	$3.74 \times 10^{-4} \pm 3.29 \times 10^{-5}$

As can be seen from Table 6, the expected (average) value for the adopted measurement is 3.74×10^{-4} m^3/s, at the result of (at the confidence level 0.99) $3.74 \times 10^{-4} \pm 3.29 \times 10^{-5}$ m^3/s, which gives an average measurement error for the analyzed series of 8.7%. The average relative mistake for the entire scope of the flow of gas amounted to ±5.3%.

4. Conclusions

Gas permeability tests of materials from UCG were carried out in bubbling conditions. This made it possible to evaluate materials with a slit-porous structure characterized by irregular shapes. The evaluation showed that the models available in the literature have a limited scope of application to skeletal media, characterized by a significant internal structure of the porous material. The test results show that under bubbling conditions, it is possible to accurately assess the gas permeability, which makes it possible to comprehensively assess the properties of the porous material in terms of the process for UCG technology. Based on the research carried out so far, it can be concluded that:

(1) Char (carbonizer) in situ and melted waste rock in situ research materials are characterized by a large variety of structures;
(2) The structural features of materials derived from underground processing can be attributed to many porosity patterns;
(3) Char (carbonizer) in situ—a fractured medium, it is more permeable in relation to melted waste rock in situ—a less-porous medium (low proportion of open pores);
(4) Average transmittance does not depend on the shape of the sample, but on the internal structure;
(5) The appropriate adaptation of the model equations allowed for a confrontation with the authors' own model, pointing to the need to interpret gas permeability in an unconventional way, especially for materials derived from UCG technology.

In the utilitarian aspect, it is possible to use the developed gas-permeability in the context of underground fermentation of coal seams with the use of an appropriate polydisperse substrate.

Funding: The study was carried out in the framework of: (1) The project under the program BIOSTRATEG founded by the National Centre for Research and Development "BIO-STRATEG1/269056/5/NCBR/2015 11 August 2015. (2) The Research Task (statutory) No. 11/79/2019 "Developing a model describing the gas permeability of anisotropic porous materials in the aspect of adhesive hydrodynamics for agroenergetic applications" implemented by the Renewable Energy Department in the Poznan Branch, Institute of Technology and Life Sciences in Falenty. The APC was funded by Institute of Technology and Life Sciences-National Research Institute, Falenty, Poland.

Institutional Review Board Statement: Not applicable.

Informed Consent Statement: Not applicable.

Data Availability Statement: The data presented in this study are available on request from the corresponding author.

Conflicts of Interest: The authors declare no conflict of interest. The funders had no role in the design of the study; in the collection, analyses, or interpretation of data; in the writing of the manuscript; or in the decision to publish the results.

Abbreviations

A	total cross-section of the flow channel	m^2
K	permeability coefficient	m^2
P	pressure gauge	Pa
Q	volumetric flow	m^3/s
Re	Reynolds number	
T	thermometer	°C
V	volume	m^3
e	indicator porosity	
d	diameter	m
f	function	
w	velocity	/s
ΔP	pressure drop, resistance flow	Pa
Ψ	tortuosity	
ε	porosity	
η	fluid viscosity	Pa·s
ξ	coefficient of flow resistance	
π	Pi number	
ρ	fluid density	kg/m^3
χ	coefficient of tortuosity	
Upper indices refer to		
a	exponent	
b	exponent	
*	own model	
Lower indices refer to		
B	acc. Brauer	
B-K	acc. Blake-Kozeny	
B-K-C	acc. Blake-Kozeny–Carman	
B-P	cc. Burke-Plummer	
E	acc. Ergun	
T	acc. Tallmadge	
V	own model	
W	acc. Windsperger	
Z	acc. Zaworonkow	
a	apparent	
b	absolute	
c	total	
ef	effective	
g	gas	
o	value calculated on the total deposit section-apparent value	
r	nozzle	
re	reference	
s	skeleton	
zm	measured	
ε	value calculated relative to the porosity	

References

1. Stańczyk, K.; Kapusta, K. Underground coal gasification. *Karbo* **2007**, *2*, 98–102.
2. Kreinnin, E.V.; Zorya, A.Y. Underground Coal Gasification Problems. *Springer Solid Fuel Chem.* **2009**, *43*, 215–218. [CrossRef]

3. Shafirovich, E.; Mastalerz, M.; Rupp, J.; Varma, A. *The Potential for Underground Coal Gasification in Indiana. Phase I. Report to the Indiana Center for Coal Technology Research (CCTR)*; Purdue University: West Lafayette, IN, USA, 2008.
4. Wałowski, G.; Filipczak, G. Technical and Technological Aspects of "In Situ" Coal Processing. In Proceedings of the XXI School of Underground Exploitation, Cracow, Poland, 20–24 February 2012; pp. 389–396.
5. Younger, P.L. Hydrogeological and Geomechanical Aspects of Underground Coal Gasification and its Direct Coupling to Carbon Capture and Storage. *Springer Mine Water Environ.* **2011**, *30*, 127–140. [CrossRef]
6. Stańczyk, K.; Świądrowski, J.; Kapusta, K.; Howaniec, N.; Cybulski, K.; Rogut, J.; Smoliński, A.; Wiatowski, M.; Kotyrba, A.; Krause, E.; et al. *Hydrogen-Oriented Underground Coal Gasification for Europe (HUGE)*; EUR 25044 EN; Final Report 2012. Research Found for Coal and Steel; European Commission: Brussels, Belgium, 2012.
7. Stańczyk, K.; Smoliński, A.; Kapusta, K.; Wiatowski, M.; Świądrowski, J.; Kotyrba, A.; Rogut, J. Dynamic experimental simulation of hydrogen oriented underground coal gasification of lignite. *Fuel* **2007**, *89*, 3307–3314. [CrossRef]
8. Wałowski, G. *Wałbrzych from the Poor Shafts to Underground Gasification of Hard Coal*; Publishing House of the Catholic Action of the Archdiocese of Lublin: Lublin, Poland, 2009; ISBN 978-83927527-8-3.
9. Wałowski, G. *Technical Condition and Prospects for the Development of Underground Coal Gasification Technology. The Influence of Young Scientists on the Achievements of Polish Science*; CREATIVETIME: Cracow, Poland, 2012.
10. Wałowski, G. *Nowa Ruda: From Poor Shafts to Underground Hard Coal Hydrogenation*; Poligraf Publishing House: Brzezia Łąka, Poland, 2011; ISBN 978-83-62752-68-3.
11. Seewald, H.; Klein, I. Methansorption an Stainkohle und Kennzeichnung der Porenstrukture. *Gluckauf Forsch.* **1985**, *47*, 149.
12. Topolnicki, J.; Wierzbicki, M.; Skoczylas, N. Tests of apparatus for measuring the kinetics of CO_2 release from hard coal samples. *Work. Strat. Mech. Inst. PAN* **2004**, *6*, 71–78.
13. Szott, W.; Gołąbek, A. Simulations of the exploitation process of natural gas deposits in shale gas formations. *Naft. Gaz* **2012**, *12*, 923–936.
14. Burdine, N.T. Relative permeability calculation from pore size distribution data. *Trans. Am. Inst. Min. Eng.* **1953**, *198*, 71–78. [CrossRef]
15. Mualem, Y. *Modeling the Hydraulic Conductivity of Unsaturated Porous Media, in Indirect Methods for Estimating the Hydraulic Properties of Unsaturated Soils*; van Genucthen, M.T., Leij, F.J., Lund, L.J., Eds.; University of California: Riverside, CA, USA, 1992; pp. 15–36.
16. Gonet, A.; Nagy, S.; Rybicki, C.; Siemek, J.; Stryczek, S.; Wiśniowski, R. Coal bed methane extraction technology (CBM). *Gor. Geol.* **2010**, *5*, 5–25.
17. Krause, E.; Karbownik, M. Tests of methane desorption and emission from samples of hard coal in the context of mine closures through flooding. *J. Sustain. Min.* **2019**, *18*, 127–133. [CrossRef]
18. Hagoort, J. *Fundamentals of Gas Reservoir Enginnering. Developments in Petroleum Science 23*; Elsevier: Amsterdam, The Netherlands, 1988.
19. Wyrwał, J. *Moisture Movement in Porous Materials and Building Partitions, WSI in Opole, Studies and Monographs*; WSI: Opole, Poland, 1989; p. 31.
20. Werner, H.; Gertis, K. *Energetische Kopplung von Feuchte- und Wärmeübertragung an Außenflächen*; Hygrische Transportphänomene in Baustoffen, Deutscher Ausschuß für Stahlbeton: Berlin, Germany, 1976; Volume 258.
21. Pogorzelski, J. *Thermal Physics of Buildings*; PWN: Warsaw, Poland, 1976.
22. Busch, A.; Gensterblum, Y.; Krooss, B.M.; Littke, R. Methane and carbon dioxide adsorption–diffusion experiments on coal: Upscaling and modeling. *Int. J. Coal Geol.* **2004**, *60*, 151–168. [CrossRef]
23. Bustin, R.M.; Bustin, A.M.M.; Cui, A.; Ross, D.; Pathi, V.M. Impact of Shale Properties on Pore Structure and Storage Characteristics. Presented at the SPE Shale Gas Production Conference, Fort Worth, TX, USA, 16–18 November 2008. Paper Number: SPE-119892-MS. [CrossRef]
24. Mastalerz, M.; He, L.; Melnichenko, Y.B.; Rupp, J.A. Porosity of coal and shale: Insights from gas adsorption and SANS/USANS techniques. *Energy Fuels* **2012**, *26*, 5109–5120. [CrossRef]
25. Karayiğit, A.İ.; Mastalerz, M.; Oskay, R.G.; Gayer, R.A. Coal petrography, mineralogy, elemental compositions and palaeoenvironmental interpretation of Late Carboniferous coal seams in three wells from the Kozlu coalfield (Zonguldak Basin, NW Turkey). *Int. J. Coal Geol.* **2018**, *187*, 54–70. [CrossRef]
26. Karayigit, A.İ.; Bircan, C.; Oskay, R.G.; Türkmen, İ.; Querol, X. The geology, mineralogy, petrography, and geochemistry of the Miocene Dursunbey coal within fluvio-lacustrine deposits, Balıkesir (Western Turkey). *Int. J. Coal Geol.* **2020**, *228*, 103548. [CrossRef]
27. Liu, X.; Nie, B.; Wang, W.; Wang, Z.; Zhang, L. The use of AFM in quantitative analysis of pore characteristics in coal and coal-bearing shale. *Mar. Pet. Geol.* **2019**, *105*, 331–337. [CrossRef]
28. Ergun, S. Fluid flow through packed columns. *Chem. Eng. Prog.* **1952**, *48*, 89–94.
29. Brauer, H. *Grundlagen der Einphasen—Und Mehrphasenströmungen*; Säuerländer: Frankfurt am Main, Germany, 1971.
30. Tallmadge, J.A. Packed bed pressure drop—An extension to higher Reynolds numbers. *AIChE J.* **1970**, *16*, 1092–1093. [CrossRef]
31. Burke, S.P.; Plummer, W.B. Gas Flow through Packed Columns. *Ind. Eng. Chem.* **1928**, *20*, 1196–1200. [CrossRef]
32. Kembłowski, Z.; Michałowski, S.; Strumiłło, C.; Zarzycki, R. *Theoretical Foundations of Chemical and Process Engineering*; WNT: Warszawa, Poland, 1985.

33. Zaworonkow, M.N. *Gidrawliczeskije Osnowy Skrubbernogo Processa i Tiepłopieredacza w Skrubberach*; Izd. Sow. Nauka: Moscow, Russia, 1944. (In Russian)
34. Windsperger, A. Abschätzung von spezifischer Oberfläche und Lückengrad bei biologischen Abluftreinigungsanlagen durch Vergleich von berechneten und experimentell erhaltenen Druckverlustwerten. *Chem. Eng. Technol.* **1991**, *63*, 80–81. [CrossRef]
35. Peszyńska, M.; Trykozko, A. Pore-to-Core simulations of flow with large velocities using continuum models and imaging data. *Comput. Geosci.* **2013**, *17*, 623–645. [CrossRef]
36. Bear, J.; Cheng, A. *Modeling Groundwater Flow and Contaminant Transport*; Springer: New York, NY, USA, 2010.
37. Wałowski, G. Model of Flow Resistance Coefficient for a Fragment of a Porous Material Deposit with Skeletal Structure. *Energies* **2021**, *14*, 3355. [CrossRef]

energies

MDPI

Article

Model-Free Control of UCG Based on Continual Optimization of Operating Variables: An Experimental Study

Ján Kačur *,†, Marek Laciak †, Milan Durdán † and Patrik Flegner †

Institute of Control and Informatization of Production Processes, Faculty BERG, Technical University of Košice, Němcovej 3, 042 00 Košice, Slovakia; marek.laciak@tuke.sk (M.L.); milan.durdan@tuke.sk (M.D.); patrik.flegner@tuke.sk (P.F.)
* Correspondence: jan.kacur@tuke.sk; Tel.: +421-55-602-5176
† These authors contributed equally to this work.

Abstract: The underground coal gasification (UCG) represents an effective coal mining technology, where coal is transformed into syngas underground. Extracted syngas is cleaned and processed for energy production. Various gasification agents can be injected into an underground georeactor, e.g., air, technical oxygen, or water steam, to ensure necessary temperature and produce syngas with the highest possible calorific value. This paper presents an experimental study where dynamic optimization of operating variables maximizes syngas calorific value during gasification. Several experiments performed on an ex situ reactor show that the optimization algorithm increased syngas calorific value. Three operation variables, i.e., airflow, oxygen flow, and syngas exhaust, were continually optimized by an algorithm of gradient method. By optimizing the manipulation variables, the calorific value of the syngas was increased by 5 MJ/m^3, both in gasification with air and additional oxygen. Furthermore, a higher average calorific value of 4.8–5.1 MJ/m^3 was achieved using supplementary oxygen. The paper describes the proposed ex situ reactor, the mathematical background of the optimization task, and results obtained during optimal control of coal gasification.

Keywords: underground coal gasification (UCG); optimization; syngas; calorific value; optimal control; operating variables; control algorithm

Citation: Kačur, J.; Laciak, M.; Durdán, M.; Flegner, P. Model-Free Control of UCG Based on Continual Optimization of Operating Variables: An Experimental Study. *Energies* **2021**, *14*, 4323. https://doi.org/10.3390/en14144323

Academic Editor: Adam Smoliński

Received: 2 June 2021
Accepted: 15 July 2021
Published: 18 July 2021

1. Introduction

The technology of underground coal gasification (UCG) enables the extraction of coal energy by thermic decomposition. The coal is transformed into syngas by the utilization of gasification agents injected into an in situ georeactor. The produced syngas is exhausted on the surface, where it is transformed into the desired form of energy, or various chemicals are produced. Compared with classical coal mining, the UCG is a less expensive technology, also attractive in terms of environmental protection. In the case of UCG, at least two boreholes must be drilled into a coal seam, i.e., inlet or injection hole and outlet production hole [1].

Before gasification can begin, a highly permeable path (i.e., channel) within the coal seam is established between the wells. This link is required as the in situ properties of the coal seam do not permit the gas flows required for economical gasification. Many of the known coal resources are currently uneconomic to mine using conventional techniques. The potential of UCG is also in the case of deposits with tectonic faults and in deposits that are unavailable for traditional mining. As coal reserves are much larger than those of natural gas, it seems likely that coal gasification will be used more frequently for generating synthesis gas to make chemicals and liquid fuels.

The essential performance parameter of coal gasification is the calorific value of the syngas.

The raw dry gas from UCG consists of hydrogen, carbon monoxide and carbon dioxide, methane, higher hydrocarbons, and traces of tars and pollutants. The valuable gases in syngas regarding calorific value are carbon monoxide, hydrogen, methane, and higher hydrocarbons [2].

The produced gas composition from UCG depends on the injected oxidant used, operating pressure, and the underground reactor's mass and energy balance [3,4].

The main chemical processes occurring during UCG are drying, pyrolysis, combustion, and solid hydrocarbon gasification. The UCG is operated as an autothermic process whereby oxygen injected through the injection borehole generates heat via the combustion reactions with the char (i.e., heterogeneous reactions of partial oxidation and combustion of carbon) and homogeneous reactions in the gas phase. The most important chemical reaction is the Boudoard chemical reaction (i.e., $C + CO_2 \rightarrow 2CO$), where valuable carbon dioxide is produced [5,6].

Various gasification agents can be injected to support gasification, but air (see [7–11]) or a mixture of oxygen and water vapor (see [7,11–13]) are most often used. It was found that the UCG technology based on injected air achieves lower syngas calorific value (approximately 6–7 MJ/m^3) as in the utilization of oxygen along with water vapor (i.e., up to 10.9 MJ/m^3, e.g., trials in Chinchilla and Rocky Mountain I [12,14]).

For the ignition of the coal seam, the pyrophosphoric materials in a gaseous state or silane-methane ignition system are used.

Figure 1 shows the UCG principle with one injection and one production well. In the underground, there is a coal seam called an in situ georeactor, in which UCG takes place. On the surface, there are devices for measurement and regulation, compressors for injection of gasifying agents, oxygen production, and devices for suction, purification, and storage of syngas.

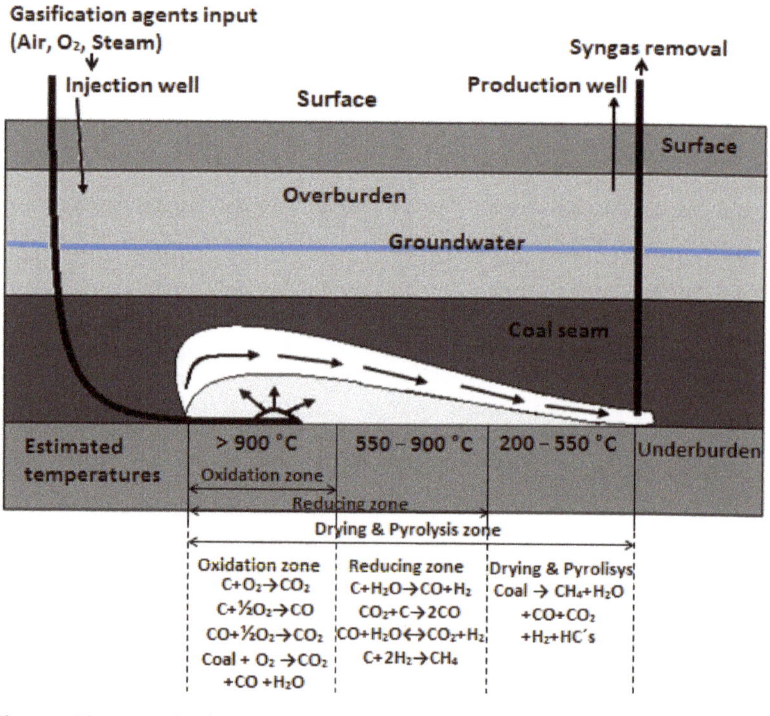

Figure 1. The principle of UCG.

The underground temperature is an essential indicator of the course of chemical reactions. Unfortunately, the measurement of underground temperature is complicated due to the inaccessibility of the measuring place. For the estimation of the underground temperature in the georeactor during UCG, various proxies are used, e.g., proxy from isotope measurement in syngas [15,16], proxy from radon emanation [17–19], or proxy based on syngas composition and machine learning model [20].

In UCG, the rock joint roughness has a significant impact on fluid migration. The surface roughness of rock directly affects its strength, deformation, and seepage characteristics. The results show that the surface roughness of sandstone increases with the increase in temperature and cycles after heating and cooling; at above 500 °C, the thermal damage increased significantly due to the expansion and cracking of quartz particles [21]. The relationship between the aperture and gas conductivity of a single natural fracture was investigated in the laboratory condition [22]. The effects of shear displacement and normal stress on the mechanical and hydraulic behavior of fractures are also studied in [23]. Many different methods have been used to measure rock fracture surface roughness. An accurate quantification of roughness is essential in modeling strength, deformability, and fluid flow behaviors of rock joints. Rock mass strength, deformability, and fluid flow behaviors depend very much on the properties of joints. A comprehensive review of rock joint roughness measurement and quantification procedures can be found in [24].

1.1. UCG Control

The main issue of UCG control is to produce syngas with stable calorific value. The control of UCG is a difficult task as, in UCG, there is a lack of direct control over many essential parameters.

The automated control of UCG can be bedeviled by a relatively considerable uncertainty of the controlled object (i.e., coal seam), which, unlike the industrial system, was created by nature. Partly this uncertainty can be reduced by a more detailed geological survey. However, even this does not guarantee the elimination of such uncertainty, as evidenced by long-term experience in the traditional coal mining technology. There may also be problems with the changing operating conditions of the process. For this reason, the control system should be robust and be able to adapt to changes in the process continuously.

The input flows of the gasifiers should be continuously optimized to ensure the required calorific value. An automated control system should be able to perform optimal control interventions and eliminate the human factor.

On the outlet from the georeactor, there is an effort to optimize suction pressure (i.e., under pressure) or directly the exhaust fan power (e.g., revolutions). The under pressure and overpressure control of UCG on the stabilization level were investigated in [25].

Kačur and Kostúr [25] have investigated the utilization of adaptive PI control to stabilize the temperature in the oxidization zone, measured by a thermocouple, with the regulated airflow and stabilization of measured oxygen concentration in syngas by regulated exhaust fan power. The adaptation of discrete controllers aimed to cope with the uncertainties in UCG. The proposed controllers were verified on an ex situ reactor.

Within experimental research of UCG improvement, an adaptive model-predictive control (AMPC) was also tested on a regression machine-based simulation model [26].

In [26], the model based on multivariate adaptive regression splines (MARS) simulated the UCG process, and the AMPC algorithm continuously optimized the three control variables (i.e., the air flow, oxygen flow, and output under pressure) to maintain the syngas calorific value setpoint. The internal prediction model of the ARX type was continuously adapted in the MPC algorithm. The simulation results showed that AMPC could achieve a better quality of calorific value stabilization with three manipulation variables than with the use of a discrete PI controller and one manipulation variable (i.e., air flow). Unfortunately, applications of MPC on industry hardware can be complicated, due to complex matrix calculations and quadratic programming [26].

Another connection of learning methods with predictive control of gasification can be found in [27]. In this work, adaptive predictive control of oxygen concentration in syngas without using a model was proposed.

Another research of automated UCG control was based on a continual adaptation of two regression models to maintain syngas calorific value [28].

This approach was also tested on an ex situ UCG reactor. Several criteria were proposed for the adaptation of regression parameters, e.g., the measure based on the desired range of the calorific value and underground temperature. The proposed models calculate optimal air flow and oxygen flow injected to the georeactor to increase or maintain the syngas calorific value during UCG operation. The regression modes were adapted by continually measured process data and the least square method. The proposed control approach has shown exciting results and demonstrated the possibility of its application on devices of industrial automation, or as a supporting algorithm for the monitoring system [28].

Another approach for estimating an optimal amount of gasification agents based on thermodynamical model was proposed in [29].

The optimization goal in [29] was to find the optimal amount of oxidizers at a known thermodynamic temperature. Optimization problems were solved by a modified gradient method for the defined weight of coal. The aim of the thermodynamics model was based on input data of UCG processes to calculate the composition of syngas at different temperatures. Based on thermodynamics, the system will be in a state of equilibrium if overall Gibson energy is at a minimal. The optimization task was solved using the method of the Lagrange multipliers [30,31].

Wei and Liu [32] have proposed a new data-based iterative optimal learning control scheme for discrete-time nonlinear systems, using the iterative adaptive dynamic programming (ADP) approach, and apply this proposed scheme to solve a coal gasification optimal tracking control problem. The neural network was used to approximate the system model using the input-state-output data of the system, and the optimal tracking control problem was transformed into a two-person zero-sum optimal regulation control problem. An iterative ADP algorithm was then established to obtain the optimal control law where the approximation errors in each iteration were considered [32].

Much research work to developing UCG advanced control has been completed by Uppal et al. [33,34].

In [34], the nonlinear time-domain UCG model was used in a closed-loop configuration with a sliding mode controller (SMC). The controller can find optimal input of the model as the molar flow rate of inlet gas (i.e., a mixture of air and steam) to keep the calorific value at the desired value in the presence of external disturbance (i.e., water influx from the surrounding aquifers). Recently, a one dimensional (1-D) packed bed model of UCG was proposed, which can be used in a closed-loop configuration with a robust controller to maintain a desired heating value of the exit gas mixture by manipulating the flow rate of injected gases. The solution of the model showed that the heating value of the exit gas is sensitive to the flow rate of inlet gases. Therefore, a robust control strategy can be employed to maintain the desired heating value in the presence of disturbances and model uncertainties by manipulating the flow rate [33,35–38].

1.2. Model-Free Control

Model-free control does not rely on any mathematical model of the controlled system. The control algorithm uses only online measurements from the real system. Model-free control can adapt to a regulated system and deal with any uncertainty in the system, as it does not rely on any specific model. One of the best-known approaches to system management without a model is extremum seeking control [39–42].

This is a model-free optimization method that was initially proposed to control train systems [43]. The main goal of such a control system is to seek an extremum, i.e., maximize or minimize a given objective function, without closed-form knowledge of the function or its gradient. In the literature, many results from the application of extremum-seeking algorithms can be found, especially after the appearance of a consistent convergence analysis in [44].

To understanding extremum seeking methods, the following general dynamics of the system can be defined:

$$\dot{x} = f(x, u), \tag{1}$$

where $x \in \mathbb{R}^n$ is the state of the system, $u \in \mathbb{R}$ is the scalar control (for simplicity), and f: $\mathbb{R}^n \times \mathbb{R} \to \mathbb{R}^n$ is a smooth function. We then use Equation (1) to represent the model of a real system, with the control goal to optimize a given performance of the system. This performance can be as simple as a regulation of a given output of the system to a desired constant value, or a more involved output tracking of the desired time-varying trajectory, etc. Let us now model this desired performance as a smooth function (x, u): $\mathbb{R}^n \times \mathbb{R} \to \mathbb{R}$, which can be simply denoted as $J(u)$, as the state vector x is driven by u.

Indeed, one of the simplest ways to maximize J is to use a gradient-based extremum seeking control as follows:

$$\dot{u} = h\frac{dJ}{du}. \tag{2}$$

Another well-known approach to seeking extremes is the so-called extremes seeking based on perturbations. The control algorithm uses the perturbation signal to examine the control space, and sets the manipulation variables toward the local optimum by implicitly monitoring the gradient update. These types of extremum search algorithms have been thoroughly analyzed, for example, in [41,44–46].

The following well-known model-free control method is the co-called reinforcement learning algorithm [47–53].

In this method, the control algorithm tests random control actions, monitors system responses, and gradually creates a system's predictive model. This approach to control belongs to the machine learning class. The algorithm continuously learns how to map system states to action interventions so that the objective function is maximized. The control algorithm creates a database of the best action interventions using trial and error [54].

In this work, a model-free optimal control design is presented, which continuously searches for local extremes of syngas calorific value using perturbation signals. The UCG represents numerous chemical processes, which complex model would be complicated. In this study, the UCG was considered a black-box system with known inputs and outputs (see Figure 2). Utilization of an ex situ reactor and the physical model of coal seam enables us to verify a UCG control system in lab-scale without environmental impact. The black-box approach and model-free control enable us to process control without the need to model internal processes. The controlled system, i.e., UCG georeactor, has known inputs and outputs. After ignition of the coal seam, the syngas production depends on injected oxidizers (i.e., gasification agents) and regulated under pressure.

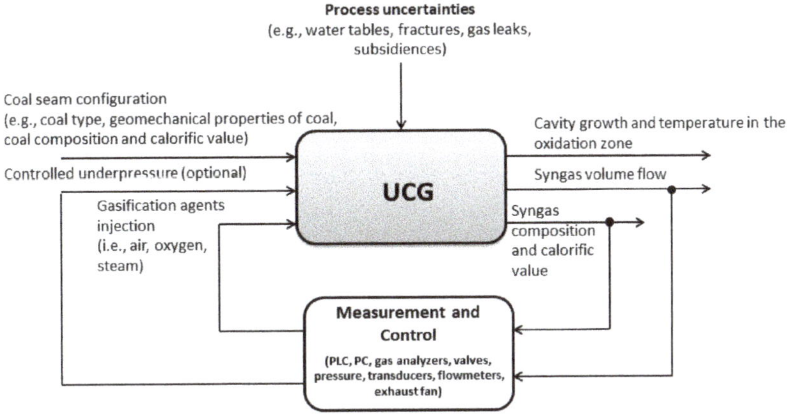

Figure 2. The UCG system configuration.

In the following sections, the UCG ex situ reactor and the proposal of model-free optimal control of UCG are described. The aim was to design a model-free control algorithm that could be implemented on industrial hardware (e.g., PLC) or SCADA software. By introducing perturbations on the controlled system, it is possible to continuously optimize the manipulation variables to maximize the calorific value of the syngas produced. The algorithm should only work with measurable inputs (i.e., airflow or oxygen flow) and measurable outputs (i.e., syngas composition and calorific value). Verification of the control system was performed on an ex situ reactor where the influence of UCG uncertainties (e.g., fractures, surface subsidies, gas leaks, and water tables, etc.) on syngas production was not analyzed at this stage of the research.

2. Experimental UCG in Ex Situ Reactor

Much research work in UCG has been performed on ex situ reactors (e.g., [6,29,55–58]). Using the ex situ reactors, the influence of various oxidants on the calorific value of syngas, the movement of the combustion front, reverse combustion, and the modeling of the temperature field was investigated. Ex situ reactors with the bedded coal represent physical models of the real coal seam. The coal storage, as well as the reactor design, meet the conditions of geometric similarity.

To test advanced control methods (i.e., adaptive and optimal), the experimental gasification equipment with a UCG ex situ reactor was constructed (see the scheme of equipment in Figure 3 and ex situ reactor in Figure 4a). A ramp with sampling points for gas analysis was constructed in the middle of the ex situ reactor. Coal was bedded longitudinally on both sides of the reactor or only on the right half.

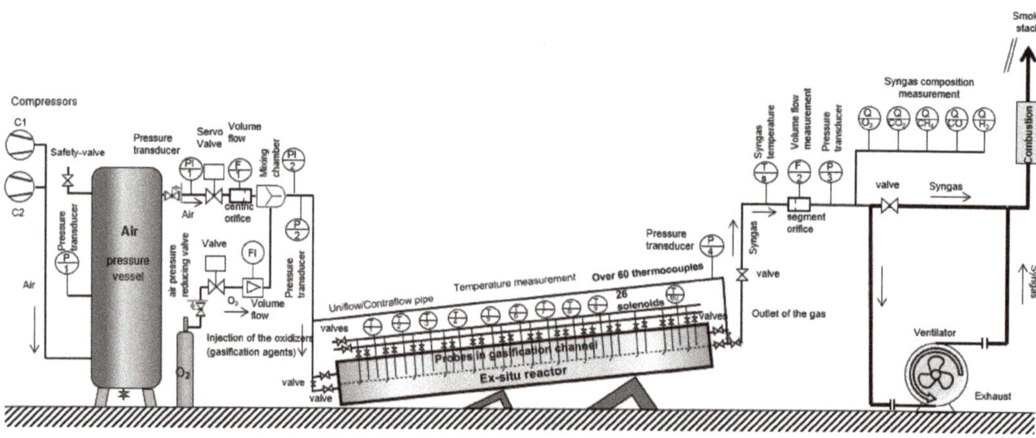

Figure 3. Experimental gasification equipment (modified after [20]).

The ex situ reactor allowed the bedding of connected blocks of coal. The coal seam model was created from glued coal blocks (see Figure 4b). For experiments, the lignite from the mine Cigel' was used. This mine belongs to Upper Nitra coal Basin in Slovakia. During the gasification, the fired coal and layers of coal from roof stone stroke down into the formed space (see Figure 4c). After gasification, there was ash in the reactor and a partly gasified coal (see Figure 4d).

Along with the coal seam model, the gasification channel was created. This channel was created as a gap between blocks or drilled as a borehole to coal blocks. The bedded coal was isolated by a mixture of sand and water glass. The outer isolation was ensured by the sibral.

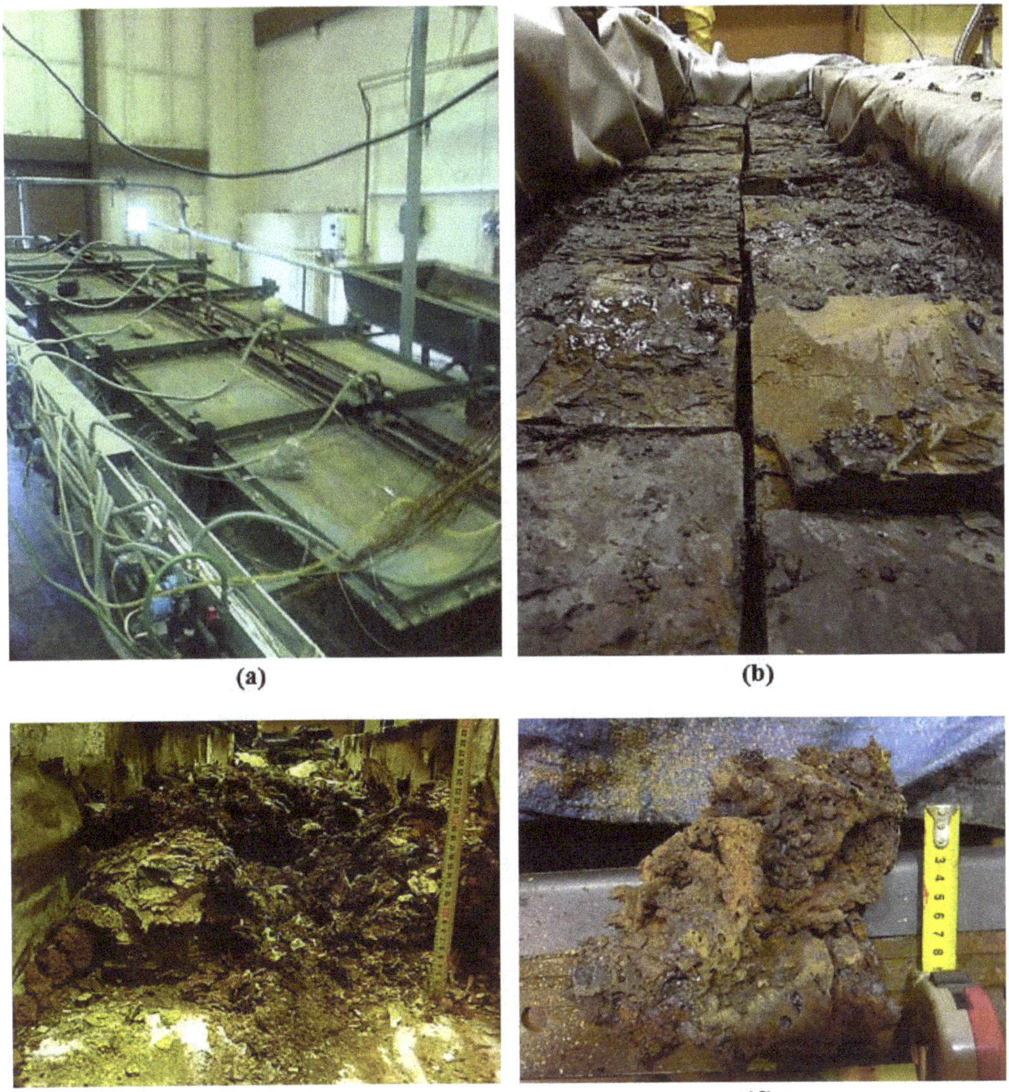

Figure 4. Ex situ reactor: (**a**) closed reactor before experiment start, (**b**) coal bedding, (**c**) gasified coal in reactor, and (**d**) extracted coal specimen.

As gasification agents, atmospheric air and technical oxygen were used. They were injected into the reactor from the pressure vessels. Two compressors produced the pressure air, and pressure oxygen was delivered from the oxygen plant.

For the measurement and control, several devices were used. Figure 5 shows devices essential for the control system verification. The volume flows of gasification agents were regulated by the servo-valve and reduction valves (see Figure 5a). Pressure transducers measured the pressure of gasification agents (see Figure 5b). The volume flows of air and syngas were measured using the flow meter diaphragms and the differential pressure transducers (see Figure 5c,h). A vortex flowmeter measured the volume flow of oxygen (see Figure 5i). The mixture of air and oxygen was created in the mixing chamber. The stationary analyzers measured the composition of syngas at the output from the ex situ reactor. Using two analyzers (i.e., Madur CMS–7 and ABB Caldos (see Figure 5d)), gas concentrations as carbon monoxide (CO_2), carbon dioxide (CO), oxygen (O_2), hydrogen (H_2), and methane (CH_4) were continually measured.

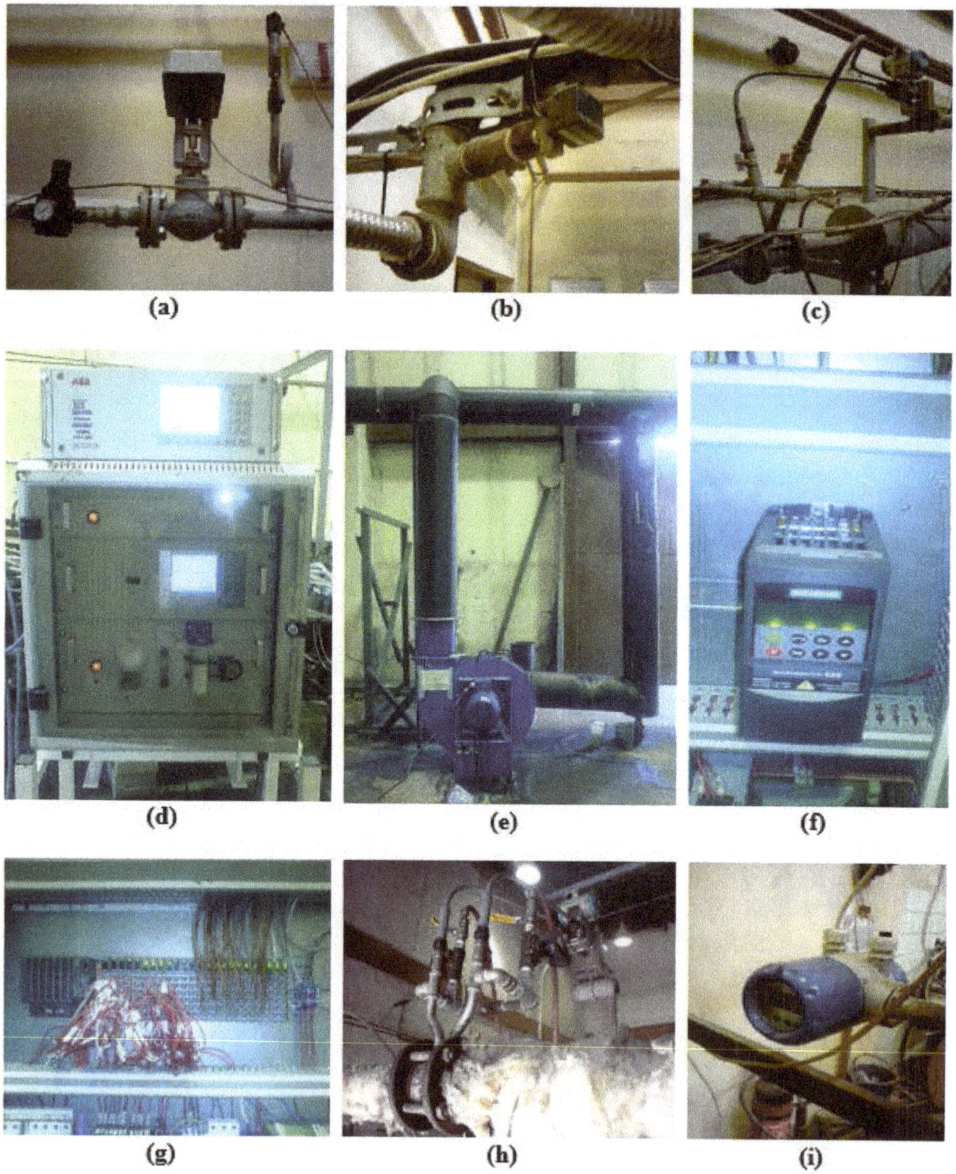

Figure 5. Devices for measurement and control: (**a**) servo valve, (**b**) pressure transducer for air, (**c**) mixing chamber and differential pressure transducer for airflow measurement, (**d**) gas analyzers, (**e**) exhaust ventilator valve, (**f**) frequency changer, (**g**) programmable logic controller (PLC), (**h**) pressure transducer for syngas flow measurement, and (**i**) vortex flow meter for oxygen flow measurement.

The syngas calorific value was calculated from the measured gases according to the following equations:

$$H = 126.4\varphi_{CO} + 107.6\varphi_{H_2} + 358.0\varphi_{CH_4}, \quad (3)$$

where H is the syngas calorific or heating value (KJ/m^3) and φ_i is the volume fraction of the i-th fuel component in percentages.

The outlet under pressure was controlled by the power of the industry fan and frequency converter (see Figure 5e,f). The produced syngas was burned in the combustion chamber or exhausted by the smoke stack.

All devices were connected to PLC (see Figure 5g). The detailed specifications of all devices used in experimental UCG can be found in [25].

The automated control system can simultaneously perform several cyclic tasks, which provide the following operations:

- control of air pressure by two compressors;
- stabilization of airflow through servo valve;
- stabilization of temperature in the oxidizing zone;
- stabilization of the O_2 concentration in syngas.

On the pipes were placed pressure transducers. The construction of an ex situ reactor enables measurement of the pressures and syngas composition, along with the physical model of the coal seam.

The temperatures in the coal, gasification channel, and isolating overburden were measured by thermocouples of K type (up to 1300 °C). All devices for the measurement and control were connected to PLC (B & R X20: Eggelsberg, Austria) that was placed along with a frequency changer to the instrumental box (i.e., switchboard).

The control algorithms (i.e., stabilization and optimal control) were implemented on PLC as cyclic tasks. The control interventions, controllers' setup, or monitoring of the gasification process variables were ensured by the SCADA system created in Promotic [2,20,25,26,28,55].

Table 1 presents the results of the coal composition analysis performed in a certified laboratory. It can be seen from the table that this is coal with higher moisture.

Table 1. Coal composition analysis (*r* = received, *d* = dry, *daf* = dry ash-free, and *a* = analytical) [25].

Parameter	Value
Total Moisture W_t^r (%)	22.25
Ash A^d (%)	26.33
Volatiles V^{daf} (%)	60.39
Carbon C^{daf} (%)	64.79
Hydrogen H^{daf} (%)	5.59
Nitrogen N^{daf} (%)	1.04
Calorific value Q_i^{daf} (MJ/kg)	24.94
Calorific value Q_i^d (MJ/kg)	18.37
Calorific value Q_i^r (MJ/kg)	13.74
Ash A^r (%)	20.47
Carbon C^r (%)	37.11
Hydrogen H^r (%)	3.20
Nitrogen N^r(%)	0.59
CaO (%)	1.12
MgO (%)	0.62

Table 1. *Cont.*

Parameter	Value
SiO_2 (%)	12.10
Al_2O_3 (%)	5.26
Fe_2O_3 (%)	2.89
Na_2O (%)	0.14
P_2O_5 (%)	0.02
TiO_2 (%)	0.17
K_2O	0.55
Volatiles V^r (%)	34.59
Analytical Moisture W^a (%)	9.56
Total Sulphur S_t^r (%)	1.93
Sulphate Sulphur S_S^r (%)	0.01
Pyritic Sulphur S_P^r (%)	1.35
Organic Sulphur S_O^r (%)	0.57
Oxygen Q^{daf} (%)	26.34
Oxygen Q^d (%)	19.4

Within the research, the permeability of the coal sample was also analyzed using a pressure permeameter. The amount of gas forced through the sample was measured with a wet gas meter. In the pressure of 1 MPa and flow 0.000011722 $m^3 \cdot s^{-1}$, the permeability coefficient had the value of $K = 2.37911 \times 10^{-14}$ m^2 [59].

3. UCG Optimal Control Based on Dynamic Optimization

An optimal control algorithm finds the optimal values of control variables (i.e., manipulation variables) to achieve a certain optimality criterion (e.g., maximum profit, minimum cost, maximum calorific value, the maximum volume of produced syngas, etc.). There are two types of automatic optimization [60]:

- Optimization with the mathematical model of the process;
- Optimization without the mathematical model of the process (i.e., the system is considered as the "black-box").

In the first case, a mathematical description of the process in the control computer is encoded. With this description, the optimal values of control variables, which are transferred to the local automatic control systems, are found. These controllers' role is to maintain the setpoint of control variables, but the system properties change during the time. Capturing the time variability in the mathematical model is probably impossible. The mathematical model consists of a wide range of parameters determined by experimental knowledge about the system's behavior.

In the second case, the optimal mode can be found via "experiment" on the observed object. In the controlled process, perturbations through the control variables are artificially created and, based on the analysis of results, are gradually improving the operating mode of equipment. The nonlinear programming methods are the most commonly used to find the optimal mode where the model is the physical object itself. The advantage of the first approach is that it involves a relatively simple calculation and setting of optimal conditions. The advantage of the second approach is that there is no required mathematical model of the process.

The disadvantage of the first approach is that obtaining a good model requires substantial theoretical and experimental efforts. The drawback of the second procedure is a break in the continual work of equipment.

To formulate optimization problems in general, so that it covers all possible cases, is very difficult. What follows is the formulation, which is quite near to the general formulation.

The definition of an optimization problem from a mathematical term is essentially a setup of set D $\in \mathbb{R}$, and a function of multiple variables $J(u_1, u_2, \ldots, u_n)$, respectively

$J(\overline{u}) \in \mathbb{R}$. Then, looking such $J(\overline{u}^{opt})$ so as $\overline{u}^{opt} \in D$, $J(\overline{u}^{opt}) \geq J(\overline{u})$ (i.e., maximization) for all $\overline{u} \in D$ was fulfilled. The set of D is often called an area created by limitations.

In terms of proposal of the optimal control system, the following approaches can be applied:

- Optimal control systems with feedback;
- Optimal control systems with feed-forward;
- Combined optimal control systems.

In this paper, to realize the optimal control of UCG, the optimal feedback control was proposed. The proposal is based on a static optimization method that was adapted to the dynamic process of UCG. The optimization method continually seeks local extremes of the objective function and optimizes manipulation variables. The advantage of this system is that it does not need the process model. In this case, the optimal control is achieved directly by the experimental way, where the model is a controlled object. Process behavior is performed in an iterative way so that the criterion of optimality according to an optimization algorithm was fulfilled. The first successful trial with optimal feedback control of UCG without model was reported in [42]. The optimal control system was based on direct extremum seeking of carbon monoxide concentration in syngas and gasification agent optimization. When formulating optimal control problems, it is desirable to know the optimized vector, optimality criterion, constraints, and have chosen a suitable optimization method.

3.1. Optimized Vector

Optimized vector (4) consists of variables that can be changed to obtain extremum of optimality criterion, which optimize these parameters. For optimization of the gasification process, this vector consists of three manipulation variables, with which we can dynamically influence the process.

$$\overline{u} = \begin{pmatrix} u_1 \\ u_2 \\ u_3 \end{pmatrix}, \tag{4}$$

where u_1 is the servo valve opening adjusted by digital pulses. By changing the servo valve opening, the change of airflow is reached (m^3/h). The airflow is the primary operating variable in experimental UCG. Although the airflow can be stabilized by PI controller, for more straightforward implementation and verification of the proposed control algorithm, we decided not to use cascade connection. Variable u_2 represents the oxygen flow added to the oxidation mixture (m^3/h). It is the value of the desired flow adjusted directly by the servo valve. The third manipulation variable, u_3, is an exhaust fan motor power frequency (Hz). The change of this variable changes the fan's revolutions, and subsequently, the suction pressure (Pa) at the outlet is affected.

3.2. Optimimality Criterion

The gasification control can be performed optimally under the chosen optimality criterion respecting restrictive conditions. Optimality systems generate optimal control, which depends on the selected optimality criterion. This criterion may be technical or economic nature. For optimal control, the gasification process can be used the following criterion of optimality:

An optimality criterion that represents the maximum calorific value of syngas was defined. The criterion was defined as functional in the form:

$$J = \int_{\tau_1}^{\tau_2} H(\tau) d\tau \rightarrow MAX \tag{5}$$

where $H(\tau)$ is the calorific value of the syngas in time τ (MJ/m^3), and constants τ_1 and τ_2 are the times of the start and the end of the analyzed section (s), respectively.

Considering that the boundaries τ_1, τ_1 in previous integral are fixed values, and time section $\tau_2 - \tau_1$ is the constant, this functional is unsuitable for practical solutions. For the realization of the optimal control system, it is preferable to use the functional in the following form:

$$J = \frac{1}{\tau_2 - \tau_1} \int_{\tau_1}^{\tau_2} H(\tau) d\tau \rightarrow MAX. \quad (6)$$

This criterion expresses the maximal calorific value on average during optimal process control.

3.3. Constraints

In formulating the optimal control problem, consideration of limitations resulting from the process is required. These are constraints and requirements on the input and output process variables, which must not be exceeded. In technological processes, such restrictions may result either from technological requirements or the equipment's design parameters, in which technological process occurs. For optimal control of the gasification process, which is performed on a laboratory gasifier, the following constraints can be defined:

- For the control variables, the constraints are defined as the following:

$$u_j^{Min} \leq u_j \leq u_j^{Max}; \quad (j = 1,\ 2,\ 3) \quad (7)$$

  where u_1^{Max}, u_1^{Min} is the maximum and minimum of allowable servo valve opening (%, pulses) or airflow (m^3/h), respectively, u_2^{Max}, u_2^{Min} is the maximum and minimum of allowable oxygen flow (m^3/h), respectively, and u_3^{Max}, u_3^{Min} represent the maximum and minimum of allowable exhaust fan power frequency (Hz), or boundaries for controlled under pressure (Pa) on the outlet from the ex situ reactor, respectively;
- If the concentration of oxygen in the syngas is too high, it means that input is set up to a high flow of oxygen or a higher amount of oxidant is blown. High oxygen concentration at the outlet leads to a surplus of oxygen in the gasification process. It is reflected in a reduced calorific value. An ideal situation occurs when the oxygen concentration on the outlet is maintained at 0%. Given the above remarks, we can define a limit on the concentration of oxygen (φ_{O_2}) in the following form:

$$\varphi_{O_2}^{Min} \leq \varphi_{O_2} \leq \varphi_{O_2}^{Max}, \quad (8)$$

  where $\varphi_{O_2}^{Max}$ is the maximal permitted concentration of O_2 in syngas (%), and $\varphi_{O_2}^{Min}$ is the minimal permitted concentration of O_2 in syngas (%).

3.4. Optimization Method

In this paper, the simple gradient optimization method with constraints is proposed to be applied for the model-free optimal control of UCG. It is an iterative method, based on the last point approximation looking for another point in which the objective function's value is nearer to the extreme. The speed of convergence of this method is quite good at greater distances from the extreme, and approaching the extreme, the rate of convergence decreases. This approach is suitable for multidimensional optimization problems and belongs to a group of methods called point–direction–step. Applications of gradient methods for solving a specific problem of optimal control can be found in [61,62].

The control algorithm aims to maintain the extreme (i.e., maximum) of the objective function in each control step. The research was aimed to maximize syngas calorific value, as this variable is the most critical indicator of UCG. Replacing the integral in Equation (6) by the operator of the summation, a discrete form of the objective function can be obtained:

$$J_k(\overline{u}) \approx \frac{1}{n}\sum_{i=1}^{n} H_j \cdot \Delta\tau_j \rightarrow MAX, \tag{9}$$

where $J_k(\overline{u})$ is the value of the objective function in the step k (MJ/m^3), k is the index of the control period $T_{0,opt}$ of the optimal control algorithm, $\overline{u}$ is the vector of the manipulation variables, which are optimized by optimal control algorithm, H is the calorific value of syngas (MJ/m^3), j is the index of the sampling period $T_{0,stab}$ on the stabilization level, $\Delta\tau_j$ is the value of $T_{0,opt}$ in the j–th step of the sampling, and n is the number of the samples in the buffer.

Equation (9) represents a nonlinear function that will be maximized. This task can be solved by a nonlinear optimization method with constraints.

Assuming that $\Delta\tau_j = 1$ for $j = 1, 2, \ldots, n$ substituting into Equation (9), then the equation for calculation of the average calorific value is the following:

$$H_k(\overline{u})^{average} = \frac{1}{n}\sum_{i=1}^{n} H_j, \tag{10}$$

where $H_k(\overline{u})^{average}$ is the average calorific value (MJ/m^3) of the syngas in the step k, and H_j is the j–th calorific value in the buffer (MJ/m^3).

The average calorific value of the produced gas during the period $T_{0,opt}$ is in the algorithm of optimal control calculated as a moving average of the samples recorded with a period $T_{0,stab}$. For calculation of the moving average, a FIFO buffer was used [63]. This buffer contains a historical record of the calorific value. The time length of this record is determined by the buffer size n and the sampling period $T_{0,stab}$.

Components of the vector $\overline{u}$ are continually optimized by optimal control algorithm to gradually achieve extremes of the objective function $J(\overline{u})$. The control system sets a new opening of the servo valve according to the value of the parameter u_1, the desired flow of the oxygen by the parameter u_2, and the new frequency of changer according to the parameter u_3. Concerning the principle of the gradient method for maximizing the objective function, the following equation is calculated iteratively [62,64]:

$$\overline{u}^{i+1} = \overline{u}^{i} + h \cdot \nabla J\left(\overline{u}^{i}\right), \tag{11}$$

where $\overline{u}^{i+1}$ is the vector of optimized control variables in step $i+1$, $\overline{u}^{i}$ is the vector of the optimized variables in the step i, and h is an iterative constant (step), which is chosen in such a way that the values $\overline{u}^{i+1}$ do not distort conditions of the existence of functions $J(\overline{u})$, also that was true (maximizing) $J\left(\overline{u}^{i+1}\right) > J\left(\overline{u}^{i}\right)$ and, at the same time, ensure appropriate convergence of the method. Step h is reduced only if the value of the objective function does not grow. Each variable of vector $\overline{u}^{i}$ can have custom parameter h. Then, the vector $\overline{h} = (h_1, h_2, h_3)^T$ can be defined; $\nabla J\left(\overline{u}^{i}\right)$ is the gradient of the objective function expressed as the column vector of partial differentials of the objective function according to variables u_1^i, u_2^i, u_3^i.

Optimization starts with the initial vector $\overline{u}^0 = \left(u_1^0, u_2^0, u_3^0\right)^T$. In the next step, the control history is perturbed, and the partial derivatives are calculated on each iteration. This straightforward control adjustment scheme derives the control perturbation for the $(i+1)$st iteration from the control gradient computed on the i-th iteration [62].

The gradient of the objective function contains the following components:

$$\nabla J\left(\overline{u}^{i}\right) = \left(\frac{\partial J}{\partial u_1^i}, \frac{\partial J}{\partial u_2^i}, \frac{\partial J}{\partial u_3^i}\right)^T. \tag{12}$$

For the approximation of gradients $\frac{\partial J}{\partial u_1^i}$, $\frac{\partial J}{\partial u_2^i}$, and $\frac{\partial J}{\partial u_3^i}$, the finite difference method can be used. Using a forward approximation, the following equations can be obtained:

$$\begin{aligned}
\frac{\partial J}{\partial u_1^i} &\cong \frac{\Delta J}{\Delta u_1^i} = \frac{J\left(u_1^i+\Delta u_1, u_2^i, u_3^i\right) - J\left(\overline{u}^i\right)}{\Delta u_1} \\
\frac{\partial J}{\partial u_2^i} &\cong \frac{\Delta J}{\Delta u_2^i} = \frac{J\left(u_1^i, u_2^i+\Delta u_2, u_3^i\right) - J\left(\overline{u}^i\right)}{\Delta u_2} \\
\frac{\partial J}{\partial u_3^i} &\cong \frac{\Delta J}{\Delta u_3^i} = \frac{J\left(u_1^i, u_2^i, u_3^i+\Delta u_3\right) - J\left(\overline{u}^i\right)}{\Delta u_3}
\end{aligned} \tag{13}$$

where Δu_j is the elemental change of the manipulation variable u_j, $(j = 1, 2, 3)$.

For the application of the gradient method, the parameters Δu_1, Δu_2, and Δu_3 are chosen so that the change of the parameter u_j by value Δu_j, (j = 1, 2, 3) causes a change in the value of the objective function. The objective function's value enters into the optimal control algorithm only from the steady state of the process. Each subsequent step of the optimal control algorithm is performed only if the following condition is fulfilled:

$$\max_{j=1\ldots n}\left\{\frac{\left|H_k(\overline{u})^{average} - H_j\right|}{H_k(\overline{u})^{average}}\right\} < \varepsilon_{max} \tag{14}$$

where $H_k(\overline{u})^{average}$ is the average syngas calorific value of the syngas in step k, H_j is the j–th calorific value in a FIFO buffer (MJ/m^3), ε_{max} is the maximum allowable deviation from the arithmetic average (%), and parameter n is the number of samples in the FIFO buffer.

Alternatively, the test of steady state can be performed by calculating standard deviation according to the equation

$$\frac{1}{n}\sqrt{\sum_{k=1}^{n}\left(H_k(\overline{u})^{average} - H_j\right)^2} < s_{max}, \tag{15}$$

where parameter s_{max} represents the maximum allowable value of the standard deviation.

The choosing constant ε_{max} is based on the analysis of the calorific value record. There was analysis of the settling time of the calorific value and the ability to stabilize. At each step of the record, the left-hand side of the Equation (14) was evaluated. Figure 6 shows the calorific value behavior during the experiment and the relative error of the arithmetic average as the value on the left side of the Equation (14). Calorific value is changed in carrying out different control interventions (e.g., change the servo valve position, increase oxygen flow, or change the power frequency of the exhaust fan). The graphical analysis follows that the calorific value can be considered steady when the relative deviation is not more than 25%.

When testing the algorithm, the buffer keeps a historical record of the samples from the last 30 min. It is time that corresponds to the approximate stabilization time of the calorific value. Detection of the steady state of the calorific value by evaluating condition (14) should be verified not only at each step k, which is performed with the period $T_{0,opt}$, but also continually in every step of the calorific value stabilization ($T_{0,stab}$). Recorded behavior of the stabilization error can serve in the analysis of the algorithm activity. If the condition (14) is fulfilled, the Boolean variable (i.e., flag) *Steady* in the algorithm is set to the logical value *True*. The buffer size and the update period can be arbitrarily set on the control panel in the monitoring system.

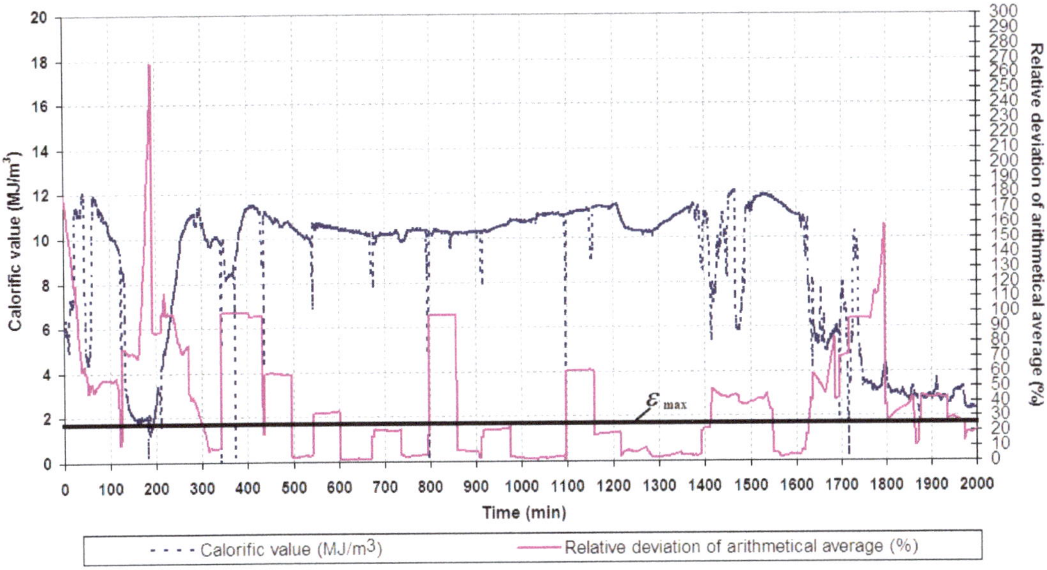

Figure 6. Principle of steady state searching.

For the automated control of UCG, the gradient method was adapted to experimental gasification equipment and possibilities of the ex situ reactor. The algorithm flow chart is shown in Figure 7. The proposed algorithm represents one of the possible variants of gradient method application for the dynamic process. The algorithm was programmed as two cyclic tasks implemented to PLC Steps of the algorithm are performed with the period $T_{0,opt}$, and this period can be arbitrarily set from the environment of the monitoring system. During testing, in the optimal control algorithm, the period $T_{0,opt}$ was set to 900 s (i.e., 15 min).

The first cyclical task worked with period $T_{0,opt}$ and ensured the implementation of each step of the algorithm, depending on the state given by flag *Steady*. The objective function's current value (in the algorithm labeled as *J*) also enters the first cyclic task.

The second cyclical task worked with the sampling period $T_{0,stab}$, detecting the stabilization of the calorific value and setting the flag *Steady*. When testing the optimal control algorithm, the period $T_{0,stab}$ was set to 1 min. The algorithm was also tested with $T_{0,stab}$ set to 30 and 15 s. At the same time, I second cyclic task ensures updating variable *J*, containing the current value of the objective function calculated according to Equation (9).

The proposed algorithm can optimize the three manipulation variables during the UCG and seek local extremes of the objective function, i.e., maximization of syngas calorific value. The algorithm uses perturbations to explore the control space, and steers the manipulation variables toward their local optimum by following a gradient update. The algorithm continuously seeks optimal manipulation variables to set the operating parameters as the airflow, the flow of additional oxygen, and the under pressure at the outlet. By automatic setup of the manipulation variables, the human factor is eliminated when deciding on action interventions to increase the syngas calorific value.

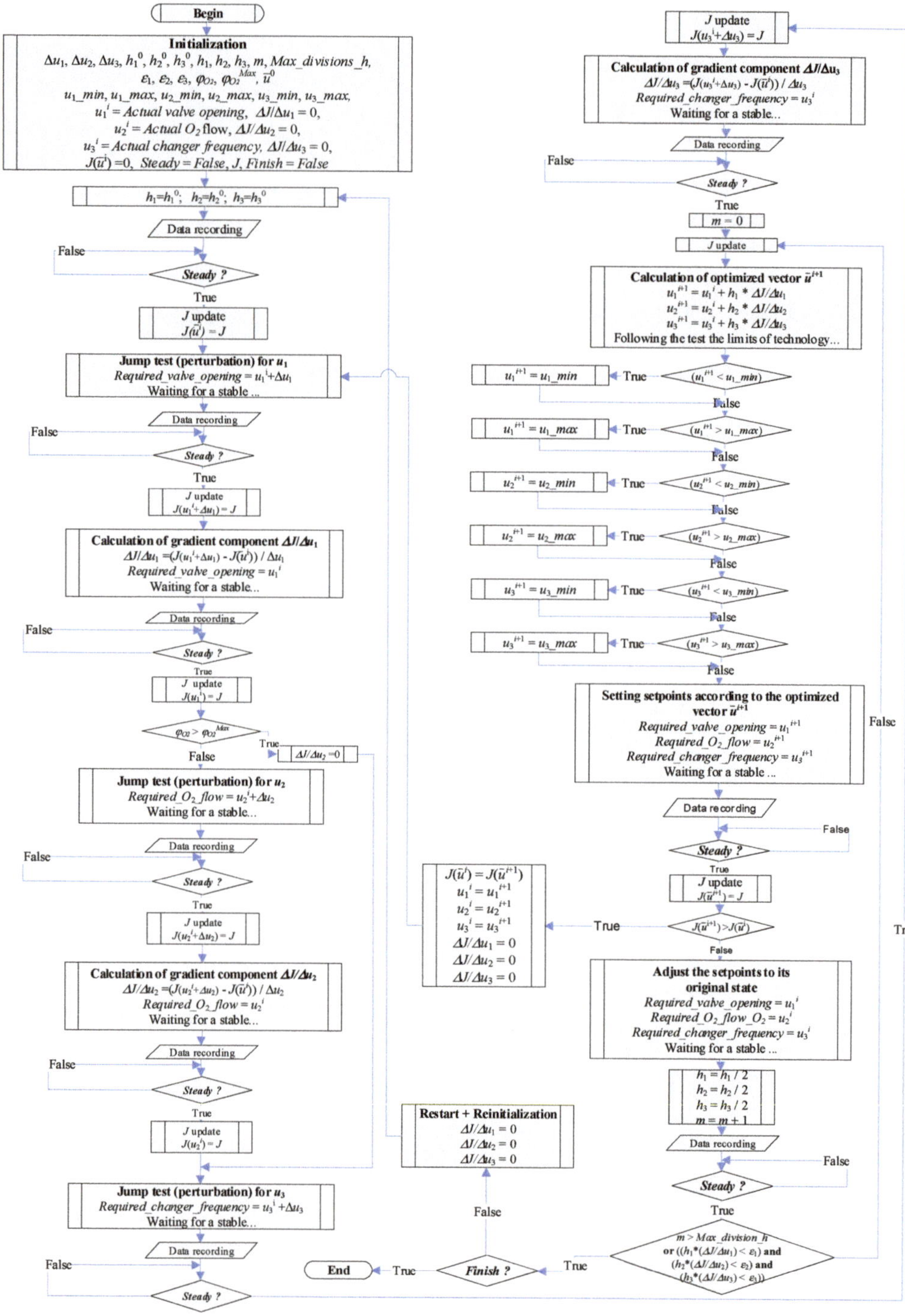

Figure 7. Algorithm of optimal control based on gradient method.

Considering a defined constraint (7), the following conditions must be evaluated in the algorithm:

$$\text{If } u_j^{i+1} > u_j^{Max} \text{ then } u_j^{i+1} = u_j^{Max}; \text{ if } u_j^{i+1} < u_j^{Min} \text{ then } u_j^{i+1} = u_j^{Min} \text{ for } j = 1, 2, 3.$$

As the proposed algorithm can set the various value of the oxygen flow rate, which in some cases can cause a state with a high excess of oxygen, the algorithm takes into account the constraint (8) and evaluates the following condition:

$$\text{If } \varphi_{O_2} > \varphi_{O_2}^{Max} \text{ then } \frac{\partial J\left(\overline{u}^i\right)}{\partial u_2^i} = 0$$

where φ_{O_2} is the currently measured concentration of O_2 in syngas and parameter $\varphi_{O_2}^{Max}$ represents the maximum allowed concentration of O_2 in syngas. When testing the algorithm, the value of the parameter $\varphi_{O_2}^{Max}$ was set to 10%.

Initialization of the algorithm requires a correctly chosen value of the optimization parameter h. The improperly selected value of this parameter can cause unexpected behavior of optimization. When the optimization is in progress, the following condition must also evaluate:

$$\text{If } J\left(\overline{u}^i\right) > J\left(\overline{u}^{i+1}\right) \text{ then } h = h/2 \text{ until } h \leq h^{Min}.$$

If $h \leq h^{Min}$, or max division of h exceeds, then it is necessary to determine the new initial vector $\overline{u}$, constant h, and re-solve the optimization task (i.e., restart and re-initiation).

4. Results

Four tests of the optimal control algorithm were performed within the experimental gasification on an ex situ reactor. The individual tests differed in the number of optimized manipulation variables and the duration. In all tests, the algorithm continuously introduced perturbations to calculate the gradients according to the Equation (13). The control algorithm optimized the manipulation variables to continuously maximize the calorific value. The algorithm worked with online mean calorific value data, which it recorded in the FIFO buffer to calculate the value of the objective function (9) within the period of optimal control $T_{0,opt}$. The algorithm was implemented on a PLC, which provided online measurement and change in manipulation variables. Changes in manipulation variables were made only when the steady state calorific value of syngas was indicated according to Equation (14). The optimal control algorithm was programmed as a cyclic task, i.e., a program implemented to PLC in Automation Basic Language. Figure 8 shows the connection diagram of the proposed UCG optimal control.

In the first experiment, the optimal control algorithm, which was tasked to maximize the calorific value and optimize two variables, was verified. Two manipulating variables, i.e., u_1—servo valve opening that adjusts the flow of injected air—and u_3—frequency of changer, that sets the exhaust fan power to effect under pressure at the outlet from the generator)—were optimized. An objective function was chosen according to Equation (9). The calorific value was entered into the algorithm as the average from the last 30 min. Within this period, the relative deviation was continuously evaluated from the arithmetic average by Equation (14). The maximum allowable relative deviation from the arithmetic average was set to 25%.

The behavior of the algorithm is shown in Figure 9. The figure shows an optimized servo valve opening to digital pulses and corresponding airflow, exhaust fan motor frequency to regulate exhausting under pressure, and behavior of calorific value with discrete values of the objective function sampled by $T_{0,opt}$. The figure also shows an improvement of the objective function concerning the value from the previous optimization step. The figure shows the application of perturbations to manipulation variables and the refinement of the parameter *h*.

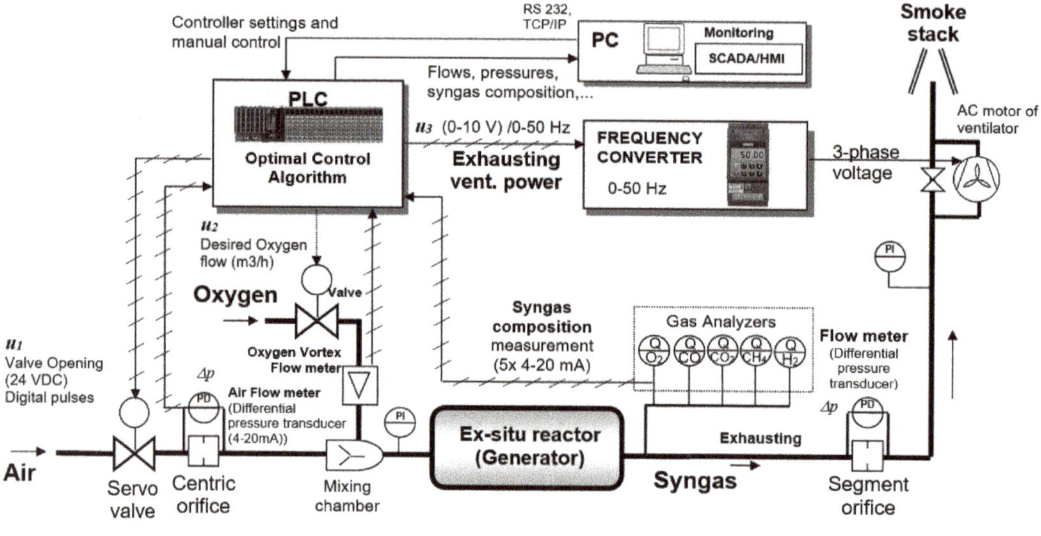

Figure 8. Scheme of the integrated control system verified in lab scale.

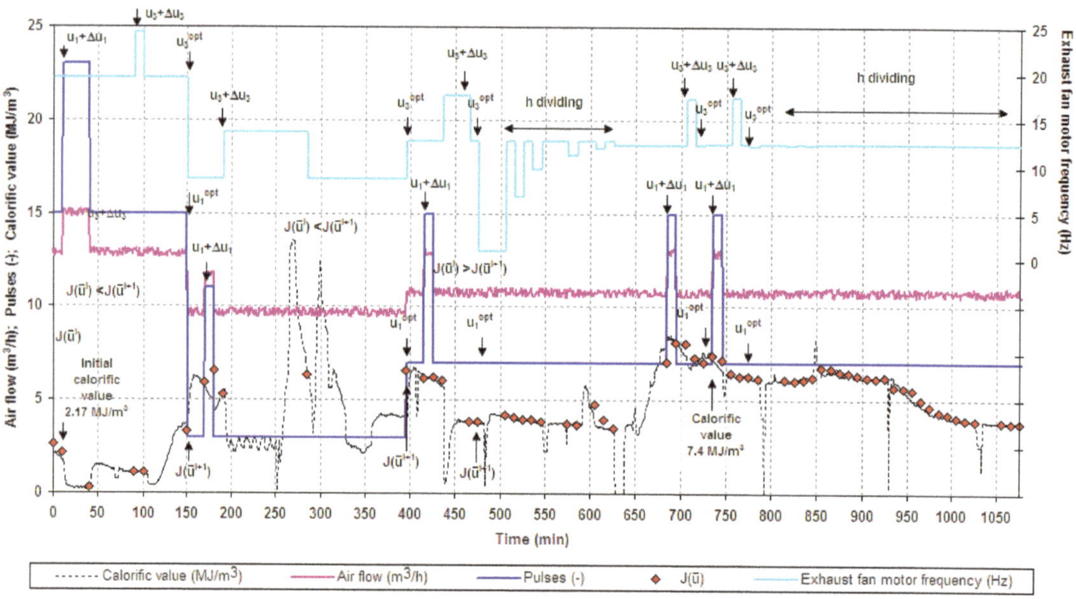

Figure 9. The first test of optimal control with optimization of two parameters.

The algorithm started with an initial value of objective function 2.17 MJ/m^3 and, after 730 min of operation, achieved an increase to 7.4 MJ/m^3. Then, calorific value only decreased, and the algorithm failed to increase it. The decrease in the calorific value was caused by low temperatures in the oxidizing zone (i.e., the temperature was less than

1000 °C). As the optimal control was deactivated after 17 h, a manual temperature increase had to be performed with the oxygen.

In another test of optimal control with optimization of two manipulation variables, the calorific value was increased from 0.95 to 2.1 MJ/m^3 (see 1025 min) (see Figure 10). The test lasted more than 9 h and, during this time, was restarted twice. The automatic restart was carried out as the algorithm failed to increase the calorific value, even after several divisions of the parameter h.

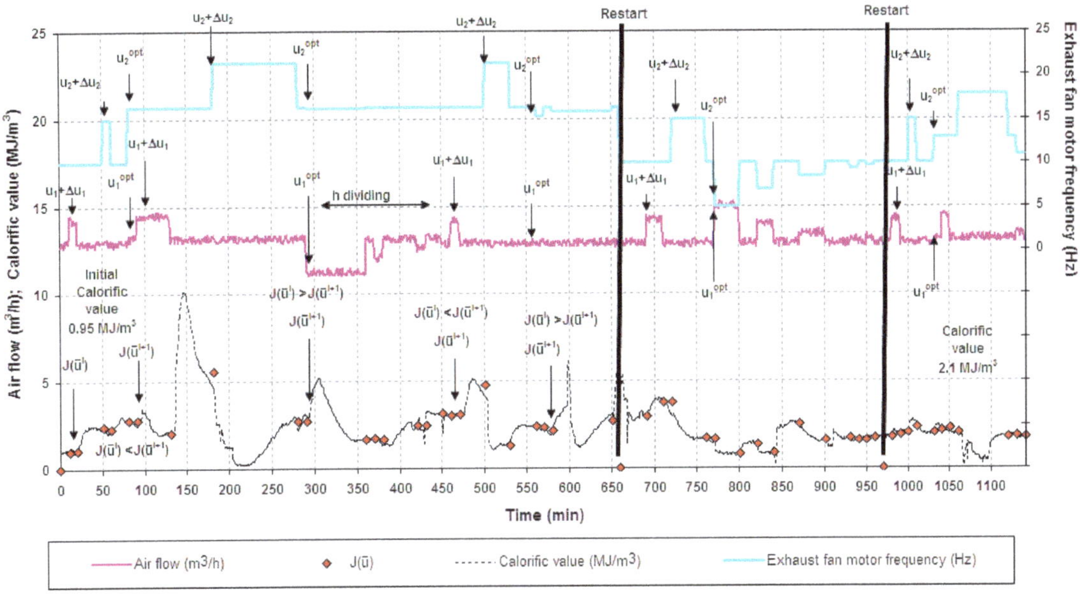

Figure 10. The second test of optimal control with optimization of two parameters.

Furthermore, an algorithm of optimal control with optimization of three variables was tested: u_1—servo valve opening, which adjusts the flow of injected air–u_2–oxygen flow—and u_3—frequency of changer, which sets the exhaust fan power. Each control variable had set its parameter h (i.e., h_1 for u_1, h_2 for u_2, and h_3 for u_3). The behavior of the algorithm activity is displayed in Figure 11. The figure shows three manipulation variables, volume flow of injected air, and the supplementary oxygen. The figure also indicates the behavior of calorific value with discrete values of the objective function, and improvement of the objective function concerning its value from the previous optimization step. The optimal control algorithm started at an initial calorific value of 4.4 MJ/m^3. During its activity for the short term, there was an increase in the calorific value to more than 9.5 MJ/m^3 (101 and 611 min). The algorithm ran a total of 10 h. During this time, it was automatically restarted twice as the dividing parameter h, in the case of unsuccessful optimization, did not bring an increase in calorific value.

In the fourth test of optimal control, three parameters were also optimized, and the calorific value of syngas was maximized. Figure 12 shows the graph presentation of the optimization activity. The algorithm ran a total of 10 h, and the calorific value increased from an initial 4.4 to 8 MJ/m^3. This result was reached by gradually reducing the airflow, increasing the flow of oxygen, and reducing exhaust fan power. The automatic restart of the algorithm was not activated during the test.

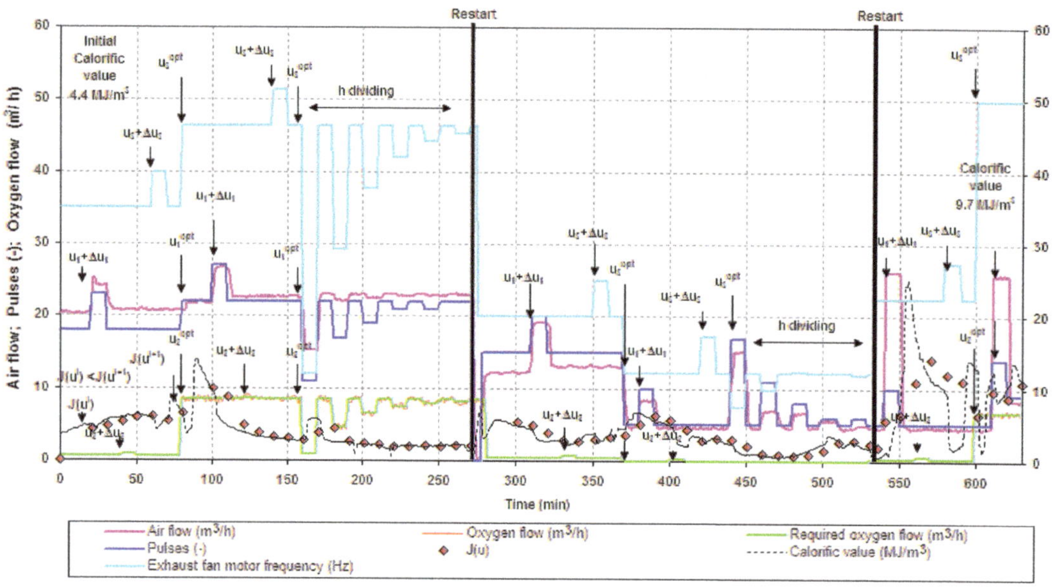

Figure 11. The first test of optimal control with optimization of three parameters.

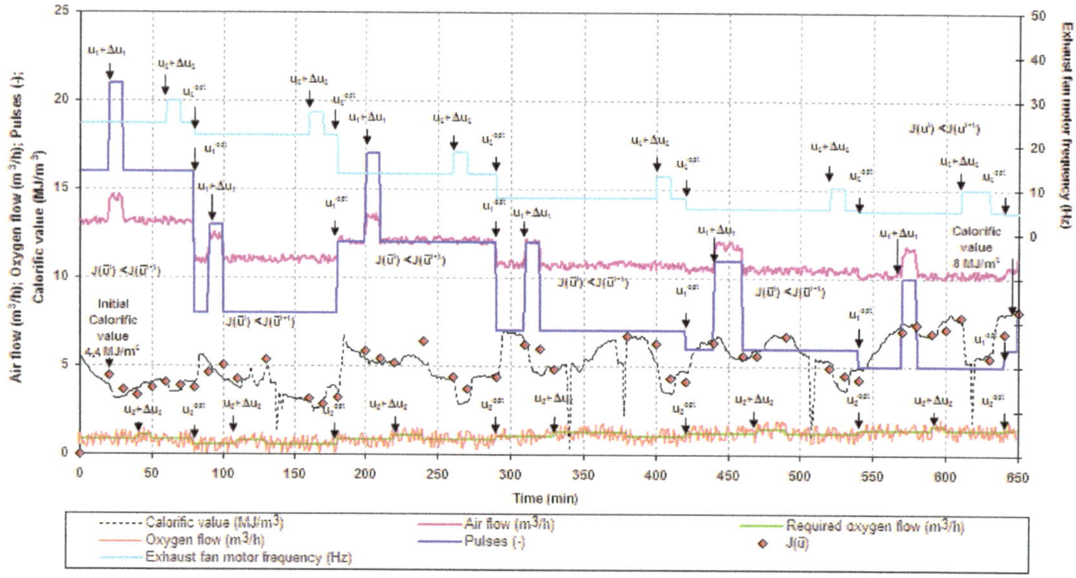

Figure 12. The second test of optimal control with optimization of three parameters.

Detailed analysis showed that the maximum permissible value of the relative deviation from the arithmetic average could also be set to a lower value for the algorithm's improvement, e.g., 10%. Smaller maximum tolerance would be extended, waiting for a stable calorific value, and extending the algorithm's test. All four tests of the algorithm favorably affect the gasification process. In any case, the calorific value was, in the short term, increased, opening the way for further improvement and testing of the algorithm.

A summary view of the configuration of individual tests and the achieved results is shown in Table 2. The UCG experiments in which the proposed control algorithm was tested differed in time duration, number of optimized manipulation variables, and achieved temperatures in the oxidation zone. The table shows that the best results, i.e., the highest calorific values, were acquired by optimizing all three manipulation variables. A higher average temperature was also obtained in these tests. The algorithm calculates manipulation variables to ensure optimal flows gasification agents so that the content of the heating components was the greatest. This aim can be achieved only at higher temperatures above 1000 °C.

Table 2. Configuration of experiments and summary of results.

Test	Duration (min)	Optimized Variables	Optimization Steps	Maximum Temperature (°C)	Average Temperature (°C)	Initial Calorific Value (MJ/m³)	Average Calorific Value (MJ/m³)	Maximum Calorific Value (MJ/m³)	Final Calorific Value (MJ/m³)
#1	1100	u_1, u_3	4	1050	878	2.17	4.3	8.1	7.4
#2	1150	u_1, u_3	5	956	815	0.95	2.1	5.5	2.1
#3	650	u_1, u_2, u_3	5	1194	938	4.4	4.8	14.2	9.7
#4	650	u_1, u_2, u_3	6	1182	1054	4.4	5.1	8.5	8.0

Optimization of supplemental oxygen injection has significantly improved gasification. On the other hand, gasification with oxygen can be more expensive.

5. Discussion

Optimal control based on the gradient method's application in the four tests ensured an increase in the calorific value. The algorithm uses perturbations to explore the control space, and steers the manipulation variable toward its local optimum by following a gradient update. Thus, the advantage of the proposed solution is that no model is needed to calculate optimal manipulation variables. However, on the other hand, the proposed control needs the continual work of equipment, and finding optimal control can be relatively time consuming. The proposed algorithm was implemented on PLC and on the SCADA system.

Three manipulation variables were optimized during experiments, i.e., airflow, oxygen flow, and exhaust fan power on the outlet. These variables enable control of the behavior of the UCG. When oxygen was used as an additional oxidant, the higher temperature in the oxidation zone and syngas' calorific value were recorded (i.e., 938–1054 °C, 4.8–5.1 MJ/m^3). There was also recorded a short-term highest calorific value of 14.2 MJ/m^3 in gasification with oxygen. When only air (i.e., combination of overpressure and under pressure system) was used as the sole oxidant, the average coal temperature and calorific value of the syngas were lower (i.e., 815–878 °C, 2.1–4.3 MJ/m^3). The highest calorific value of 8.1 MJ/m^3 was recorded during gasification with air.

Other algorithms for continuous extreme search have also been experimentally tested at UCG, and they are compared with the proposed algorithm in this paper.

Kostúr and Kačur [42] proposed an extremal controller algorithm that continuously maximized CO concentration in the syngas. The discrete extremal controller algorithm optimized only one manipulation variable, which stabilized the airflow by the PI controller. When testing the control algorithm, the CO concentration in the syngas increased by 17 vol.% (from 15 to 31%), which resulted in an increase in the calorific value of the syngas by 5 MJ/m^3 (from 12 to 15 MJ/m^3).

Kačur et al. [28] also tested optimal control with an adaptive regression model. The online adaptation of the regression model was based on the least-squares method, and the selected criterion was based on the calorific value of the syngas. The syngas calorific value was successfully maintained from 3 to 10 MJ/m^3 during the experiment. However, this algorithm was more computationally complex and implemented within the SCADA system. In both cases, UCG control tests were performed on the same type of coal and the same ex situ reactor as used in this work.

Despite the complexity of the control system, the implemented system remains open to further improvement. It is necessary to look for such a solution so that the calorific value of syngas is sustainable in the long run. For further research in laboratory conditions, it is necessary to test the optimal control algorithm for another criterion, e.g., maximizing the temperature in the oxidation zone and maximizing economic profit or CO concentration in the produced gas.

6. Conclusions

Underground coal gasification has a range of potential benefits. The process is safer and more energy-efficient than the conventional combination of coal mining and surface combustion. At present, UCG is blindly controlled, and various automated control methods are being tested experimentally.

In laboratory conditions, the technology of UCG was investigated on an ex situ reactor with a coal seam model to find a suitable control strategy. This paper solved the problem of model-free optimal control of underground coal gasification of UCG. The proposed algorithm was based on a simple gradient method that optimizes by experimental way of operating variables.

The novelty and originality of the proposed solution of UCG control rests on a model-free approach. This approach of automated UCG control to maximize calorific value has not been investigated to date. Most of the research in the world focuses on the model-based stabilization of calorific value and mathematical modeling of UCG processes. However, research is lacking in improving the direct automated control of the UCG process that is often controlled only blindly. Furthermore, the high complexity of model-based control complicates the implementation of control on automation hardware (e.g., due to the calculation of matrices or quadratic programming). The advantage of the proposed control is the low computational complexity and the possibility of implementation on industrial hardware.

The proposed control algorithm has been implemented on a PLC; it does not require a process model, and only online measured process data are needed. The proposed algorithm, in four tests, was able to increase the calorific value by optimizing the manipulation variables. Better algorithm performance, i.e., higher syngas calorific value, was achieved by optimizing three manipulation variables, i.e., when additional oxygen flow was optimized. By optimizing the manipulation variables, the calorific value of the syngas was increased by 5 MJ/m^3, both in gasification with air and additional oxygen. However, in gasification with additional oxygen, higher average calorific values (i.e., 5.1 MJ/m^3) were achieved. When gasifying with air (overpressure and under pressure), the highest average calorific value obtained was 4.3 MJ/m^3. Gasification with air showed a lower average temperature in the oxidation zone (878 °C) than gasification with additional oxygen (1054 °C).

Although the control system has only been tested on an experimental ex situ gasifier with minor modifications, it can also be tested on an in situ gasifier operation. Proposed control can eliminate the human factor when deciding on actions to intervene in the UCG process (i.e., correct setting of control parameters) when maintaining the maximum calorific value is necessary.

Author Contributions: Conceptualization, J.K. and M.L.; data curation, M.D.; formal analysis, P.F. and M.D.; methodology, J.K. and M.L.; project administration, J.K.; resources, J.K. and M.L.; supervision, M.D. and P.F.; validation, P.F.; writing—original draft preparation, J.K. and M.L.; and writing—review and editing M.D. and P.F. All authors have read and agreed to the published version of the manuscript.

Funding: This research was funded by project COGAR RFCR-CT-2013-00002, Slovak Research and Development Agency, under contract No. APVV-18-0526 and No. APVV-14-0892.

Institutional Review Board Statement: Not applicable.

Informed Consent Statement: Not applicable.

Data Availability Statement: No new data were created or analyzed in this study. Data sharing is not applicable to this article.

Acknowledgments: This work was supported by the Slovak Research and Development Agency under contract No. APVV-18-0526 and APVV-14-0892.

Conflicts of Interest: The authors declare no conflict of interest.

Abbreviations

The following abbreviations are used in this paper:

UCG	Underground coal gasification
PI	Proportional integral
MPC	Model-predictive control
AMPC	Adaptive model-predictive control
ARX	Auto-regressive with eXogenous input
MARS	Multivariate adaptive regression splines
ADP	Adaptive dynamic programming
SMC	Sliding mode control
1-D	One dimensional
3-D	Three dimensional
PLC	Programmable logic controller
SCADA	Supervisory control and data acquisition
FIFO	First in, first out

Nomenclature

x	The state of the controlled system
u	The manipulation variable
H	The syngas calorific value (MJ/m^3)
φ_{CO}	The volume fraction of CO in syngas (%)
φ_{H_2}	The volume fraction of H_2 in syngas (%)
$\overline{u}$	The volume fraction of CH4 in syngas (%)
$\overline{u}^0$	The initial vector of optimized manipulation variables
$\overline{u}^{opt}$	The vector of optimized manipulation variables
$\mathbb{R}$	Real numbers
$J(\overline{u})$	The objective function
u_1	The change of the servo valve position in pulses or airflow (m^3/h)
u_2	The volume flow of oxygen added to the oxidation mixture (m^3/h)
u_3	The exhaust fan motor power frequency (Hz) or controlled under pressure (Pa)
u_j	The j-th manipulation variable (j = 1, 2, 3)
u_j^{Min}	Minimum of the j-th manipulation variable
u_j^{Max}	Maximum of the j-th manipulation variable
u_1^{Max}, u_1^{Min}	Boundaries of servo valve opening (%, pulses) or airflow (m^3/h)
u_2^{Max}, u_2^{Min}	Boundaries of oxygen flow (m^3/h)
u_3^{Max}, u_3^{Min}	Boundaries of exhaust fan power frequency (Hz) or under pressure (Pa)
φ_{O_2}	The volume fraction of O2 in syngas (%)
$\varphi_{O_2}^{Min}$	The minimal permitted concentration of O2 in syngas (%)
$\varphi_{O_2}^{Mqx}$	The maximal permitted concentration of O2 in syngas (%)
τ	Time (s)
τ_1, τ_2	Time of beginning and end of the analyzed section (s)

K	The coefficient of permeability (m^2)
k	The index of the control period $T_{0,opt}$ of the optimal control algorithm
J_k	The value of the objective function in the step k (MJ/m^3)
$H_k(\overline{u})^{average}$	The average calorific value (MJ/m^3) of the syngas in the step k
H_j	The j–th calorific value in the buffer (MJ/m^3)
n	The number of the samples in the FIFO buffer
$\Delta\tau_j$	The value of $T_{0,stab}$ in the j–th step of the sampling
$T_{0,opt}$	The optimal control sampling period (s)
$T_{0,stab}$	The sampling period on the stabilization level (s)
$\overline{u}^i$	The vector of optimized control variables in the step i
$\overline{u}^{i+1}$	The vector of the optimized variables in the step $i+1$
h	The common iterative constant
$\overline{h}$	The vector of iterative constants for each manipulation variable
$\nabla J\left(\overline{u}^i\right)$	The gradient of the objective function
Δu_j	Increments of manipulation variables ($j = 1, 2, 3$)
ε_{mqx}	The maximum allowable deviation from the arithmetic average (%)
s_{max}	The maximum allowable value of the standard deviation

References

1. Perkins, G. *Mathematical Modelling of Underground Coal Gasification, Academic Dissertation*; University of New South Wales, Faculty of Science and Engineering: Sydney, NSW, Australia, 2005.
2. Kačur, J.; Durdán, M.; Laciak, M.; Flegner, P. Impact analysis of the oxidant in the process of underground coal gasification. *Measurement* **2014**, *51*, 147–155. [CrossRef]
3. Chandellea, V.; Lia, T.K.; Ledenta, P.; Patignya, J.; Henqueta, H.; Kowola, K.; Massona, G.; Mostadea, M. *Belgo-German Experiment on Underground Gasification Demonstration Project*; Technical Report; Commission of the European Communities: Brussels, Belgium, 1989; ISBN 92-825-9673-7.
4. Creedy, D.P.; Garner, K.; Holloway, S.; Jones, N.; Ren, T.X. *Review of Underground Coal Gasification Technological Advancements*; Technical Report COAL R11 DTI/Pub URN 01/1041; Department of Trade and Industry: London, UK, 2001.
5. Anon. *Underground Coal Gasification Program*; Report ERDA 77-51/4 on Contract No. EX-76-C-01-2343, Technical Report; Booz, Allen & Hamilton, Inc.: Washington, DC, USA; US Energy Research and Development Administration: McLean, VA, USA, 1977.
6. Dobbs, R.L.; Krantz, W.B. *Combustion front Propagation in Underground Coal Gasification, Final Report, Work Performed under Grant No. DE-FG22-86PC90512*; Technical Report; Boulder Department of Chemical Engineering, University of Colorado: Boulder, CO, USA, 1990.
7. Cena, R.J.; Thorsness, C.B. *Underground Coal Gasification Database*; Technical Report UCID-19169; Lawrence Livermore National Laboratory, University of California: Berkeley, CA, USA, 1981.
8. Gibb, A. *The Underground Gasification of Coal*; Sir Isaac Pitman & Sons Ltd.: London, UK, 1964.
9. Gregg, D.W.; Hill, R.W.; Olness, D.U. *An Overview of the Soviet Effort in Underground Coal Gasification*; Technical Report UCRL-52004; Lawrence Livermore Laboratory, University of California: Berkeley, CA, USA, 1976.
10. Olness, D.U. *The Angrenskaya Undergroud Coal Gasification Station*; Technical Report UCRL-53300; Lawrence Livermore National Laboratory, University of California: Berkeley, CA, USA, 1982.
11. Anon. *Underground Coal Gasification—First Trial in the Framework of a Community Collaboration: Final Summary Report*; Technical Report; Underground Gasification Europe (UGE): Alcorisa, Spain, 1999.
12. Boysen, J.E.; Canfield, M.T.; Covell, J.R.; Schmit, C.R. *Detailed Evaluation of Process and Environmental Data from the Rocky Mountain I Underground Coal Gasification Field Test*; Technical Report No. GRI-97/0331; Gas Research Institute: Chicago, IL, USA, 1998.
13. Robert, G.M.M.; Oliver, L.; Spackman, L.K. Field and laboratory results from the TONO 1 (CRIP) UCG Cavity Excavation Project. WIDCO Mine Site, Centralia, Washington. *Fuel Sci. Technol. Int.* **1989**, *7*, 1059–1120.
14. Walker, L.K.; Blinderman, M.S.; Brun, K. An IGCC Project at Chinchilla, Australia Based on Underground Coal Gasification UCG. In Proceedings of the 2001 Gasification Technologies Conference, San Francisco, CA, USA, 8–10 October 2001; pp. 1–6.
15. Koenen, M.; Bergen, F.; David, P. Isotope Measurements as a Proxy for Optimizing Future Hydrogen Production in Underground Coal Gasification, News in Depth. 2015. Available online: https://www.tno.nl/media/2624/information20-nid1.pdf (accessed on 26 May 2021).
16. Brasseur, A.; Antenucci, D.; Bouquegneau, J.-M.; Coëme, A.; Dauby, P.; Létolle, R.; Mostade, M.; Pirlot, P.; Pirard, J.-P. Carbon stable isotope analysis as a tool for tracing temperature during the El Tremedal underground coal gasification at great depth. *Fuel* **2002**, *81*, 109–117. [CrossRef]
17. Wu, J.M. Radon Distribution Under the Mine and the Application of Radon Measuring in the Monitoring of the Natural Fire Zone. Ph.D. Thesis, Shanxi Institute of Mining and Technology, Taiyuan, China, 1994.
18. Wu, H.S. *The Measuring Methods of Radon and Its Application*; Nuclear Energy Press: Beijing, China, 1995.

19. Wu, H.S.; Bai, Y.S.; Lin, Y.F.; Chang, G.L. The Action of Relay Transmission of Radon Migration. *Chin. J. Geophys.* **1997**, *40*, 136–142.
20. Kačur, J.; Durdán, M.; Laciak, M.; Flegner, P. A Comparative Study of Data-Driven Modeling Methods for Soft-Sensing in Underground Coal Gasification. *Acta Polytech.* **2019**, *59*, 322–351. [CrossRef]
21. Ge, Z.; Sun, Q.; Zhang, N. Changes in surface roughness of sandstone after heating and cooling cycles. *Arab. J. Geosci.* **2020**, *13*. [CrossRef]
22. Schrauf, T.W.; Evans, D.D. Laboratory studies of gas flow through a single natural fracture. *Water Resour. Res.* **1986**, *22*, 1038–1050. [CrossRef]
23. Chen, Z.; Narayan, S.P.; Yang, Z.; Rahman, S.S. An experimental investigation of hydraulic behavior of fractures and joints in granitic rock. *Int. J. Rock Mech. Min. Sci.* **2000**, *37*, 1061–1071. [CrossRef]
24. Ge, Y.; Kulatilake, P.H.S.W.; Tang, H.; Xiong, C. Investigation of natural rock joint roughness. *Comput. Geotech.* **2014**, *55*, 290–305. [CrossRef]
25. Kačur, J.; Kostúr, K. Approaches to the Gas Control in UCG. *Acta Polytech.* **2017**, *57*, 182–200. [CrossRef]
26. Kačur, J.; Flegner, P.; Durdán, M.; Laciak, M. Model Predictive Control of UCG: An Experiment and Simulation Study. *Inf. Technol. Control* **2019**, *48*, 557–578. [CrossRef]
27. Hou, Z.; Liu, S.; Yin, C. Local Learning-based Model-Free Adaptive Predictive Control for Adjustment of Oxygen Concentration in Syngas Manufacturing Industry. *IET Control Theory Appl.* **2016**, *10*, 1384–1394. [CrossRef]
28. Kačur, J.; Laciak, M.; Durdán, M.; Flegner, P. Application of multivariate adaptive regression in soft-sensing and control of UCG. *Int. J. Model. Identif. Control* **2019**, *33*, 246–260. [CrossRef]
29. Laciak, M.; Ráškayová, D. The using of thermodynamic model for the optimal setting of input parameters in the UCG process. In Proceedings of the ICCC 2016 17th International Carpathian Control Conference, High Tatras, Slovakia, 29 May–1 June 2016; pp. 418–423. [CrossRef]
30. Laciak, M.; Kostúr, K.; Durdán, M.; Kačur, J.; Flegner, P. The analysis of the underground coal gasification in experimental equipment. *Energy* **2016**, *114*, 332–343. [CrossRef]
31. Laciak, M.; Kačur, J.; Kostúr, K. The verification of thermodynamic model for UCG process. In Proceedings of the ICCC 2016 17th International Carpathian Control Conference, High Tatras, Slovakia, 29 May–1 June 2016; pp. 424–428. [CrossRef]
32. Wei, Q.; Liu, D. Adaptive dynamic programming for optimal tracking control of unknown nonlinear systems with application to coal gasification. *Trans. Autom. Sci. Eng. IEEE* **2014**, *11*, 1020–1036. [CrossRef]
33. Uppal, A.A.; Bhatti, A.I.; Aamir, E.; Samar, R.; Khan, S.A. Control oriented modeling and optimization of one dimensional packed bed model of underground coal gasification. *J. Process Control* **2014**, *24*, 269–277. [CrossRef]
34. Arshad, A.; Bhatti, A.I.; Samar, R.; Ahmed, Q.; Aamir, E. Model Development of UCG and Calorific Value Maintenance via Sliding Mode Control. In Proceedings of the 2012 International Conference on Emerging Technologies, Islamabad, Pakistan, 8–9 October 2012; pp. 1–6. [CrossRef]
35. Uppal, A.A.; Bhatti, A.; Aamir, E.; Samar, R.; Khan, S. Optimization and control of one dimensional packed bed model of underground coal gasification. *J. Process Control* **2015**, *35*, 11–20. [CrossRef]
36. Uppal, A.A.; Alsmadi, Y.M.; Utkin, V.I.; Bhatti, A.I.; Khan, S.A. Sliding mode control of underground coal gasification energy conversion process. *IEEE Trans. Control Syst. Technol.* **2018**, *26*, 587–598. [CrossRef]
37. Uppal, A.A.; Butt, S.S.; Bhatti, A.I.; Aschemann, H. Integral sliding mode control and gain-scheduled modified Utkin observer for an underground coal gasification energy conversion process. In Proceedings of the 23rd International Conference on Methods and Models in Automation and Robotics, Miedzyzdroje, Poland, 27–30 August 2018; pp. 357–362. [CrossRef]
38. Uppal, A.A.; Butt, S.S.; Khan, Q.; Aschemann, H. Robust tracking of the heating value in an underground coal gasification process using dynamic integral sliding mode control and a gain-scheduled modified Utkin observer. *J. Process Control* **2019**, *73*, 113–122. [CrossRef]
39. Sira Ramirez, H.; Luviano-Juárez, A.; Ramírez-Neria, M.; Zurita-Bustamante, E.W. *Active Disturbance Rejection Control of Dynamic Systems*; Butterworth-Heinemann an Imprint of Elsevier: Oxford, UK, 2017.
40. Zhang, C.; Ordóñez, R. *Extremum-Seeking Control and Applications: A Numerical Optimization-Based Approach*; Springer: New York, NY, USA, 2012. [CrossRef]
41. Ariyur, K.B.; Krstić, M. *Real-Time Optimization by Extremum-Seeking Control*; John Wiley & Sons, Inc.: New York, NY, USA, 2003. [CrossRef]
42. Kostúr, K.; Kačur, J. Extremum Seeking Control of Carbon Monoxide Concentration in Underground Coal Gasification. *IFAC PapersOnLine* **2017**, *50*, 13772–13777. [CrossRef]
43. Leblanc, M. Sur lélectrification des chemins de fer au moyen de courants alternatifs de fréquence élevée. *Rev. Générale Lelectricité* **1922**, *12*, 275–277.
44. Krstić, M.; Wang, H.H. Stability of extremum seeking feedback for general nonlinear dynamic systems. *Automatica* **2000**, *36*, 595–601. [CrossRef]
45. Tan, Y.; Nesic, D.; Mareels, I. On the dither choice in extremum seeking control. *Automatica* **2008**, *44*, 1446–1450. [CrossRef]
46. Rotea, M.A. Analysis of multivariable extremum seeking algorithms. In Proceedings of the 2000 American Control Conference ACC (IEEE Cat. No.00CH36334), Chicago, IL, USA, 28–30 June 2000; pp. 433–437. [CrossRef]

47. Busoniu, L.; Babuska, R.; De Schutter, B. A Comprehensive Survey of Multiagent Reinforcement Learning. *IEEE Trans. Syst. Man Cybern. Part C* **2008**, *38*, 156–172. [CrossRef]
48. Rao, R.P.N. *Reinforcement Learning: An Introduction*; Sutton, R.S., Barto, A.G., Eds.; MIT Press: Cambridge, MA, USA, 1998; 380p, ISBN 0-262-19398-1.
49. Bertsekas, D.; Tsitsiklis, J. Neuro-dynamic programming: An overview. In Proceedings of the 34th IEEE Conference on Decision and Control, New Orleans, LA, USA, 13–15 December 1995; Volume 1, pp. 560–564. [CrossRef]
50. Szepesvári, C. Algorithms for Reinforcement Learning. *Synth. Lect. Artif. Intell. Mach. Learn.* **2010**, *4*, 1–103. [CrossRef]
51. Busoniu, L.; Babuska, R.; De Schutter, B.; Ernst, D. *Reinforcement Learning and Dynamic Programming Using Function Approximators*; CRC Press: Boca Raton, FL, USA, 2010. [CrossRef]
52. Farahmand, A. Regularization in Reinforcement Learning. Ph.D. Thesis, University of Alberta, Edmonton, AB, Canada, 2011.
53. Kormushev, P.; Calinon, S.; Caldwell, D.G. Robot motor skill coordination with EM-based Reinforcement Learning. In Proceedings of the 2010 IEEE/RSJ International Conference on Intelligent Robots and Systems, Taipei, Taiwan, 18–22 October 2010; pp. 3232–3237. [CrossRef]
54. Benosman, M. Adaptive Control: An Overview. In *Learning-Based Adaptive Control*; Elsevier: New York, NY, USA, 2016; pp. 1–35. [CrossRef]
55. Kostúr, K.; Laciak, M.; Durdán, M.; Kačur, J.; Flegner, P. Low-calorific gasification of underground coal with a higher humidity. *Measurement* **2015**, *63*, 69–80. [CrossRef]
56. Wiatowski, M.; Kapusta, K.; Stańczyk, K. Efficiency assessment of underground gasification of ortho-and meta-lignite: Highpressure ex-situ experimental simulations. *Fuel* **2019**, *236*, 221–227. [CrossRef]
57. Stańczyk, K.; Kapusta, K.; Wiatowski, M.; Swiadrowski, J.; Smolinski, A.; Rogut, J. Experimental simulation of hard coal underground gasification for hydrogen production. *Fuel* **2012**, *91*, 40–50. [CrossRef]
58. Kapusta, K.; Wiatowski, M.; Stańczyk, K. An experimental ex-situ study of the suitability of a high moisture ortho-lignite for underground coal gasification (UCG) process. *Fuel* **2016**, *179*, 150–155. [CrossRef]
59. Kačur, J.; Kostúr, K. *Research Report of the Project "Underground Gasification by Thermal Decomposition No. APVV-0582-06 for Year 2008"*; Technical Report; Technical University of Košice, Faculty BERG and HBP a.s. Prievidza: Košice, Slovakia, 2008.
60. Hrubina, K. *Optimálne Riadenie I a II*; Edičné stredisko VŠT v Košiciach: Košice, Slovakia, 1987.
61. Hrubina, K. *Riešenie Problémov Optimálneho Riadenia, Metódami Numerickej Analýzy, Záverečná Správa o Fakultnej Výskumnej Úlohe*; Technical Report; E-Fak 9 VŠT Košiciach: Košice, Slovakia, 1975.
62. Stengel, R.F. *Optimal Control and Estimation*; Dover, Inc.: New York, NY, USA, 1994.
63. Hanuš, B.; Modrlák, O.; Olehla, M. *Číslicová Regulace Technologických Procesů. Algoritmy, Matematicko-Fyzikální analýza, Identifikace, Adaptace, 1. Vydání*; Vysoké Učení Technické v Brňe, VUTIUM: Brno, Czech Republic, 2000.
64. Kostúr, K. *Optimalizácia Procesov*; Edičné stredisko TU v Košiciach: Košice, Slovakia, 1991.

MDPI
St. Alban-Anlage 66
4052 Basel
Switzerland
Tel. +41 61 683 77 34
Fax +41 61 302 89 18
www.mdpi.com

Energies Editorial Office
E-mail: energies@mdpi.com
www.mdpi.com/journal/energies

www.ingramcontent.com/pod-product-compliance
Lightning Source LLC
Chambersburg PA
CBHW061607210926
51850CB00009B/2246
9783036560113